© 2016
Clement Ampadu
drampadu@hotmail.com

ISBN: 978-1-365-25998-2
ID: 19080932
www.lulu.com

All rights reserved. No part of this publication may be produced or transmitted in any form or by any means, electronic or mechanical, including photocopying and recording, or in any information storage and retrieval system, without the prior written permission of the publisher.

Contents

	Dedication	3
1	**Higher-Order Banach Contraction Mapping Theorem in Generalized Multiplicative Cone b-Metric Space**	**4**
1.1	Brief Summary	4
1.2	Preliminaries	4
1.3	Main Results	6
1.4	Exercises	7
1.5	References	8
2	**Fixed Point Theorems in Ordered Multiplicative Cone b-Metric Space with Generalized Multiplicative c-Distance**	**9**
2.1	Brief Summary	9
2.2	Preliminaries	9
2.3	Main Results	12
2.4	Exercises	15
2.5	References	17
3	**r-Points of Coincidence and Common r-Fixed Points for Higher-Order Expansive Type Multiplicative Contractions in Multiplicative Cone b-Metric Space**	**18**
3.1	Brief Summary	18
3.2	Preliminaries	18
3.3	Main Results	21
3.4	Exercises	23
3.5	References	23
4	**Common r-Fixed Point Theorems in b-TVS Multiplicative Cone Metric Space**	**24**
4.1	Brief Summary	24
4.2	Preliminaries	24
4.3	Main Results	26
4.4	Exercises	28
4.5	References	29
5	**r-Coincidence Point and r-Fixed Point Theorems in Multiplicative Cone b-Metric Space**	**30**
5.1	Brief Summary	30
5.2	Preliminaries	30
5.3	Main Results	33
5.4	Exercises	36
5.5	References	37
6	**r-g-Monotone and r-w-Compatible Mappings in Ordered Multiplicative Cone b-Metric Space**	**38**
6.1	Brief Summary	38
6.2	Preliminaries	38
6.3	Main Result	40
6.4	Exercises	43
6.5	References	44

Dedication

This book is dedicated to those who read it .

Clement Ampadu
August, 2016

Chapter 1

Higher-Order Banach Contraction Mapping Theorem in Generalized Multiplicative Cone b-Metric Space

1.1 Brief Summary

> **Abstract A.1 1**
>
> In this chapter we introduce a concept of generalized multiplicative cone b-metric space which generalize concepts of multiplicative metric space, rectangular multiplicative metric space, multiplicative b-metric space, multiplicative cone metric space, multiplicative cone rectangular metric space and multiplicative cone b-metric space. An analogue of the higher-order Banach contraction principle [Ampadu, Clement (2015):Generalization of Higher Order Contraction Mapping Theorem, Unpublished] is also proved.

1.2 Preliminaries

Multiplicative cone in soft metric spaces was given by Ampadu [Ampadu,Clement (2015): Multiplicative Soft Cone Metric Spaces and Some Fixed Point Theorems for Multiplicative Expanding Mappings, Unpublished; Ampadu, Clement (2015): Multiplicative Soft Cone Metric Spaces and Some Fixed Point Theorems for Multiplicative Contraction Mappings, Unpublished]. It follows from these papers that we have the following

> **Definition A.1 1**
>
> Let E be a real Banach space and P be a subset of E. We will say P is a *multiplicative cone* if
>
> (a) P is closed, nonempty, and satisfies $P \neq \{1\}$
>
> (b) $x^a \cdot y^b \in P$ for all $x, y \in X$ and non-negative real numbers a, b
>
> (c) $x \in P$ and $\frac{1}{x} \in P$ imply $x = 1$, that is, $P \cap \frac{1}{P} = 1$

> **Definition A.2 1**
>
> Given a multiplicative cone $P \subset E$, we define a partial ordering $\preceq$ with respect to P by $x \preceq y$ iff $\frac{y}{x} \in P$. We will write $x < y$ if $x \preceq y$ and $x \neq y$, and $x \ll y$ if $\frac{y}{x} \in int\ P$, where $int\ P$ denotes the interior of P. A cone P will be called a solid cone if $int\ P \neq \emptyset$

CHAPTER 1. HIGHER-ORDER BANACH CONTRACTION MAPPING THEOREM IN GENERALIZED MULTIPLICATIVE CONE B-METRIC SPACE

Definition A.3 1

Let X be a nonempty set. Suppose that the mapping $m : X \times X \mapsto E$ satisfies

(a) $1 \preceq m(x,y)$ for all $x, y \in X$ and $m(x,y) = 1$ iff $x = y$

(b) $m(x,y) = m(y,x)$ for all $x, y \in X$

(c) $m(x,y) \preceq m(x,z) \cdot m(z,y)$ for all $x, y, z \in X$

Then m will be called a multiplicative cone metric on X and (X, m) will be called a multiplicative cone metric space

Definition A.4 1

Let X be a nonempty set and $s \geq 1$ be a real number. Suppose that the mapping $m : X \times X \mapsto E$ satisfies

(a) $1 \preceq m(x,y)$ for all $x, y \in X$ and $m(x,y) = 1$ iff $x = y$

(b) $m(x,y) = m(y,x)$ for all $x, y \in X$

(c) $m(x,y) \preceq [m(x,z) \cdot m(z,y)]^s$ for all $x, y, z \in X$

Then m will be called multiplicative cone b-metric on X and (X, m) will be called a multiplicative cone b-metric space

Definition A.5 1

Let X be a nonempty set. Suppose that the mapping $m : X \times X \mapsto E$ satisfies

(a) $1 \preceq m(x,y)$ for all $x, y \in X$ and $m(x,y) = 1$ iff $x = y$

(b) $m(x,y) = m(y,x)$ for all $x, y \in X$

(c) $m(x,y) \preceq m(x,u) \cdot m(u,v) \cdot m(v,y)$ for all $x, y \in X$ and for all distinct points $u, v \in X - \{x, y\}$

Then m will be a called a multiplicative cone rectangular metric on X and (X, m) a multiplicative cone rectangular metric space.

Definition A.6 1

Let X be a nonempty set and $s \geq 1$ be a real number. Suppose that the mapping $m : X \times X \mapsto E$ satisfies

(a) $1 \preceq m(x,y)$ for all $x, y \in X$ and $m(x,y) = 1$ iff $x = y$

(b) $m(x,y) = m(y,x)$ for all $x, y \in X$

(c) $m(x,y) \preceq [m(x,u) \cdot m(u,v) \cdot m(v,y)]^s$ for all $x, y \in X$ and for all distinct points $u, v \in X - \{x, y\}$

Then m will be a called a generalized multiplicative cone b-metric on X and (X, m) a generalized multiplicative cone b-metric space.

Example A.7 1

Let $E = \mathbb{R}^2$, $P = \{(x,y) \in E | x, y \geq 1\}$, $X = A \cup B$, where $A = \{\frac{1}{n} : n \in \mathbb{N}\}$ and $B = \mathbb{N}$. Set $a > 1$ and define $m : X \times X \mapsto E$ such that $m(x,y) = m(y,x)$ for all $x, y \in X$ and $m(x,y) = (1,1)$, if $x = y$; $m(x,y) = (a^2, a^2)$, if $x, y \in A$; $m(x,y) = (a^{\frac{1}{2n}}, a^{\frac{1}{2n}})$, if $x = \frac{1}{n} \in A$ and $y \in \{2, 3\}$; $m(x,y) = (a, a)$, otherwise. Then (X, m) is a generalized multiplicative cone b-metric space with coefficent $s = 2 > 1$

> **Example A.8 1**
>
> Let $E = \mathbb{R}^2$, $P = \{(x,y) \in E | x, y \geq 1\}$, $X = \mathbb{N}$. Set $a > 1$ and let $m : X \times X \mapsto E$ be such that $m(x,y) = (1,1)$, for all $x = y \in X$; $m(x,y) = m(y,x)$ for all $x, y \in X$; $m(x,y) = (a^{10}, a^{10})$, if $x = 1, y = 2$; $m(x,y) = (a,a)$, if $x \in \{1,2\}$ and $y = 3$; $m(x,y) = (a^2, a^2)$, if $x \in \{1,2,3\}$ and $y = 4$; $m(x,y) = (a^3, a^3)$, if x or $y \notin \{1,2,3,4\}$ and $x \neq y$. Then (X, m) is a generalized multiplicative cone b-metric space

> **Definition A.9 1**
>
> For any $x \in X$ we define the multiplicative open ball with center x and radius $r > 1$ by $B_r(x) = \{y \in X : m(x,y) < r\}$

> **Remark A.10 1**
>
> Let U be the collection of all subsets A of X satisfying the condition that for each $x \in A$ there exists $r > 1$ such that $B_r(x) \subseteq A$. Then U defines a topology for the generalized multiplicative cone b-metric space, (X, m)

> **Definition A.11 1**
>
> Let (X, m) be a generalized multiplicative cone b-metric space. The sequence $\{x_n\}$ will be called
>
> (a) a multiplicative convergent sequence if for every $c \in E$ with $1 \ll c$, there is $n_0 \in \mathbb{N}$ such that for all $n \geq n_0$, $m(x_n, x) \ll c$ for some $x \in X$
>
> (b) a multiplicative Cauchy sequence if for all $c \in E$ with $1 \ll c$, there is $n_0 \in \mathbb{N}$ such that for all $n, k \geq n_0$, $m(x_n, x_k) \ll c$

> **Definition A.12 1**
>
> The generalized multiplicative cone b-metric space (X, m) will be called multiplicative complete if every multiplicative Cauchy sequence in X is multiplicative convergent in X

1.3 Main Results

> **Theorem A.1 1**
>
> Let (X, m) be a complete generalized multiplicative cone b-metric space with coefficient $s > 1$, P be a solid cone and $T : X \mapsto X$ be a mapping satisfying $m(T^r x, T^r y) \leq m(x,y)^{Zq^r}$ for all $x, y \in X$, where $Z \geq 1$ and $q \in [0, \frac{1}{s})$ are given by Proposition 1.11 [Ampadu, Clement (2016): Higher Order Banach Contraction Principle in Rectangular Multiplicative b-Metric Space, Unpublished]. Then T has a unique fixed point

> **Proof of Theorem A.1 1**
>
> Let $x_0 \in X$ be arbitrary. Define a sequence $\{x_n\}$ by $x_{n+1} = T^r x_n$ for all $n \geq 0$. We show that $\{x_n\}$ is a multiplicative Cauchy sequence. If $x_n = x_{n+1}$ then x_n is a r-fixed point of T. So we assume that $x_n \neq x_{n+1}$ for all $n \geq 0$. Set $m(x_n, x_{n+1}) = m_n$, it follows that $m(x_n, x_{n+1}) = m(T^r x_{n-1}, T^r x_n) \preceq m(x_{n-1}, x_n)^{Zq^r}$, that is, $m_n \preceq m_{n-1}^{Zq^r}$, and by induction we obtain $m_n \preceq m_0^{(Zq^r)^n}$. Further we assume that x_0 is not a r-periodic point of T. Indeed if $x_0 = x_n$, then for any $n \geq 2$, we have, $m_0 = m(x_0, x_1) = m(x_0, T^r x_0) = m(x_n, T^r x_n) = m(x_n, x_{n+1}) = m_n \preceq m_0^{(Zq^r)^n}$, and since $Zq^r < 1$, then $\frac{1}{m_0} \in P$, therefore, $m_0 = 1$, that is $x_0 = x_1$, and so x_0 is a r-fixed point of T. Thus, we assume that $x_n \neq x_k$ for all distinct $n, k \in \mathbb{N}$. Now put $m(x_n, x_{n+2}) = m_n^*$, it follows that $m(x_n, x_{n+2}) = m(T^r x_{n-1}, T^r x_{n+1}) \preceq m(x_{n-1}, x_{n+1})^{Zq^r}$, that is, $m_n^* \preceq (m_{n-1}^*)^{Zq^r}$, and by induction we obtain $m_n^* \preceq (m_0^*)^{(Zq^r)^n}$. For the obtained sequence $\{x_n\}$ we consider two possible cases for $m(x_n, x_{n+p})$. If p is odd, say $p = 2k+1$, then we deduce that $m(x_n, x_{n+2k+1}) \preceq m_0^{\frac{1+Zq^r}{1-s(Zq^r)^2} s(Zq^r)^n}$. Let $1 \ll c$. Choose $\delta > 1$ such that $\frac{c}{N_\delta(1)} \subseteq P$, where $N_\delta(1) = \{y \in E : \|y\| < \delta\}$. Also choose a natural number N_1 such that $m_0^{\frac{1+Zq^r}{1-s(Zq^r)^2} s(Zq^r)^n} \in N_\delta(1)$ for all $n \geq N_1$. Then, $m_0^{\frac{1+Zq^r}{1-s(Zq^r)^2} s(Zq^r)^n} \ll c$ for all $n \geq N_1$. Thus, $m(x_n, x_{n+2k+1}) \preceq m_0^{\frac{1+Zq^r}{1-s(Zq^r)^2} s(Zq^r)^n} \ll c$ for all $n \geq N_1$. If p is even, say $p = 2k$, then we deduce that $m(x_n, x_{n+2k}) \preceq m_0^{\frac{1+Zq^r}{1-s(Zq^r)^2} s(Zq^r)^n} \cdot (m_0^*)^{(Zq^r)^{n-2}}$. Now choosing a natural number N_2 such that $m_0^{\frac{1+Zq^r}{1-s(Zq^r)^2} s(Zq^r)^n} \cdot (m_0^*)^{(Zq^r)^{n-2}} \in N_\delta(1)$, for all $n \geq N_2$, then, $m_0^{\frac{1+Zq^r}{1-s(Zq^r)^2} s(Zq^r)^n} \cdot (m_0^*)^{(Zq^r)^{n-2}} \ll c$ for all $n \geq N_2$, thus, $m(x_n, x_{n+2k}) \preceq m_0^{\frac{1+Zq^r}{1-s(Zq^r)^2} s(Zq^r)^n} \cdot (m_0^*)^{(Zq^r)^{n-2}} \ll c$ for all $n \geq N_2$. Now let $N_0 = \max\{N_1, N_2\}$, then for all $n \geq N_0$, we have $m(x_n, x_{n+p}) \ll c$ as $n \to \infty$. Thus, $\{x_n\}$ is a multiplicative Cauchy sequence in X. By the multiplicative completeness of (X, m), there exists $u \in X$ such that $x_n \to u$ as $n \to \infty$. We now show u is a r-fixed point of T. For any $n \in \mathbb{N}$, we have
>
> $$\begin{aligned} m(u, T^r u) &\preceq [m(u, x_n) \cdot m(x_n, x_{n+1}) \cdot m(x_{n+1}, T^r u)]^s \\ &= [m(u, x_n) \cdot m_n \cdot m(T^r x_n, T^r u)]^s \\ &\preceq [m(u, x_n) \cdot m_n \cdot m(x_n, u)^{Zq^r}]^s \\ &\preceq [m(x_n, u)^{1+Zq^r} \cdot m_0^{(Zq^r)^n}]^s \end{aligned}$$
>
> Now choose N_3, N_4 such that $m(x_n, u) \ll c^{\frac{1}{2s(1+Zq^r)}}$ for all $n \geq N_3$ and $m_0^{(Zq^r)^n} \ll c^{\frac{1}{2s}}$ for all $n \geq N_4$ and let $N_0 = \max\{N_3, N_4\}$. Then for all $n \geq N_0$, $m(u, T^r u) \ll c$. It follows that $m(u, T^r u) = 1$, that is, $u = T^r u$. Thus, u is a r-fixed point of T. For uniqueness, let v be another r-fixed point of T, then it follows that $m(u, v) = m(T^r u, T^r v) \leq m(u, v)^{Zq^r}$, but $1 - Zq^r \neq 0$, thus, $m(u, v) = 1$, that is, $u = v$ and uniqueness follows.

1.4 Exercises

> **Exercise A.1 1**
>
> Verify Example A.7

> **Exercise A.2 1**
>
> Consider the proof of Theorem A.1. Verify the following inequality by mathematical induction, $m_n^* \preceq (m_0^*)^{(Zq^r)^n}$

Exercise A.3 1

Consider the proof of Theorem A.1, explain how one deduces the following

(a) $m(x_n, x_{n+2k+1}) \preceq m_0^{\frac{1+Zq^r}{1-s(Zq^r)^2}s(Zq^r)^n}$

(b) $m(x_n, x_{n+2k}) \preceq m_0^{\frac{1+Zq^r}{1-s(Zq^r)^2}s(Zq^r)^n} \cdot (m_0^*)^{(Zq^r)^{n-2}}$

Hint: Reny George et.al, GENERALIZED CONE b-METRIC SPACES AND CONTRACTION PRINCIPLES, MATEMATIQKI VESNIK 67, 4 (2015), 246–257

Verifying Multiplicative b-Metric Space 1

Obtain the higher-order version of the Kannan contraction principle in generalized multiplicative cone b-metric space

Hint: See Theorem 3.3 of Reny George et.al, GENERALIZED CONE b-METRIC SPACES AND CONTRACTION PRINCIPLES, MATEMATIQKI VESNIK 67, 4 (2015), 246–257

1.5 References

(1) Ampadu, Clement (2015):Generalization of Higher Order Contraction Mapping Theorem, Unpublished

(2) Ampadu, Clement (2015): Multiplicative Soft Cone Metric Spaces and Some Fixed Point Theorems for Multiplicative Expanding Mappings, Unpublished

(3) Ampadu, Clement (2015): Multiplicative Soft Cone Metric Spaces and Some Fixed Point Theorems for Multiplicative Contraction Mappings, Unpublished

(4) Ampadu, Clement (2016): Higher Order Banach Contraction Principle in Rectangular Multiplicative b-Metric Space, Unpublished

(5) Reny George et.al, GENERALIZED CONE b-METRIC SPACES AND CONTRACTION PRINCIPLES, MATEMATIQKI VESNIK 67, 4 (2015), 246–257

Chapter 2

Fixed Point Theorems in Ordered Multiplicative Cone b-Metric Space with Generalized Multiplicative c-Distance

2.1 Brief Summary

Abstract B.1 1

In this chapter we introduce a concept of generalized multiplicative c-distance which can be considered a multiplicative generalization of c-distance which already exist in the literature. We prove some theorems for contractive mappings in multiplicative cone b-metric space by using the generalized multiplicative c-distance. Some examples are given to strengthen the main results.

2.2 Preliminaries

Multiplicative cone in soft metric spaces was given in [Ampadu,Clement (2015): Multiplicative Soft Cone Metric Spaces and Some Fixed Point Theorems for Multiplicative Expanding Mappings, Unpublished; Ampadu, Clement (2015): Multiplicative Soft Cone Metric Spaces and Some Fixed Point Theorems for Multiplicative Contraction Mappings, Unpublished]. It follows from these papers that we have the following

Definition B.1 1

Let E be a real Banach space and P be a subset of E. We will say P is a *multiplicative cone* if

(a) P is closed, nonempty, and satisfies $P \neq \{1\}$

(b) $x^a \cdot y^b \in P$ for all $x, y \in X$ and non-negative real numbers a, b

(c) $x \in P$ and $\frac{1}{x} \in P$ imply $x = 1$, that is, $P \cap \frac{1}{P} = 1$

Definition B.2 1

Given a multiplicative cone $P \subset E$, we define a partial ordering $\preceq$ with respect to P by $x \preceq y$ iff $\frac{y}{x} \in P$. We will write $x < y$ if $x \preceq y$ and $x \neq y$, and $x \ll y$ if $\frac{y}{x} \in int\ P$, where $int\ P$ denotes the interior of P. A multiplicative cone P will be multiplicative normal if there is a number $N > 0$ such that for all $x, y \in P$, $1 \leq x \leq y$ implies $\|x\| \leq \|y\|^N$. The least positive number satisfying the inequality, $\|x\| \leq \|y\|^N$, will be called the multiplicative normal constant of P

Definition B.3 1

Let X be a nonempty set. Suppose that the mapping $m : X \times X \mapsto E$ satisfies

(a) $1 \preceq m(x, y)$ for all $x, y \in X$ and $m(x, y) = 1$ iff $x = y$

(b) $m(x, y) = m(y, x)$ for all $x, y \in X$

(c) $m(x, y) \preceq m(x, z) \cdot m(z, y)$ for all $x, y, z \in X$

Then m will be called a multiplicative cone metric on X and (X, m) will be called a multiplicative cone metric space

Definition B.4 1

Let X be a nonempty set and $s \geq 1$ be a real number. Suppose that the mapping $m : X \times X \mapsto E$ satisfies

(a) $1 \preceq m(x, y)$ for all $x, y \in X$ and $m(x, y) = 1$ iff $x = y$

(b) $m(x, y) = m(y, x)$ for all $x, y \in X$

(c) $m(x, y) \preceq [m(x, z) \cdot m(z, y)]^s$ for all $x, y, z \in X$

Then m will be called multiplicative cone b-metric on X and (X, m) will be called a multiplicative cone b-metric space

Definition B.5 1

Let (X, m) be a cone b-metric space. The sequence $\{x_n\}$ will be called

(a) a multiplicative convergent sequence if for every $c \in E$ with $1 \ll c$, there is $n_0 \in \mathbb{N}$ such that for all $n \geq n_0$, $m(x_n, x) \ll c$ for some $x \in X$

(b) a multiplicative Cauchy sequence if for all $c \in E$ with $1 \ll c$, there is $n_0 \in \mathbb{N}$ such that for all $n, k \geq n_0$, $m(x_n, x_k) \ll c$

Definition B.6 1

The cone b-metric space (X, m) will be called multiplicative complete if every multiplicative Cauchy sequence in X is multiplicative convergent in X

Remark B.7 1

Let E be a real Banach space with multiplicative cone P

(a) If $\alpha \leq \alpha^\lambda$, where $\alpha \in P$ and $0 < \lambda < 1$, then $\alpha = 1$

(b) If $c \in int\ P$, $a_n \to 1$, as $n \to \infty$. Then there exists a positive integer N such that $a_n \ll c$ for all $n \geq N$

Definition B.8 1

Let (X, m) be a multiplicative cone metric space, then a function $q_m : X \times X \mapsto E$ will be called a multiplicative c-distance on X if the following conditions are satisfied

(a) $1 \leq q_m(x, y)$ for all $x, y \in X$

(b) $q_m(x, y) \leq q_m(x, z) \cdot q_m(z, y)$ for all $x, y, z \in X$

(c) for each $x \in X$ and $n \geq 1$, if $q_m(x, y_n) \leq u$ for some $u = u_x \in P$, then, $q_m(x, y) \leq u$ whenever $\{y_n\}$ is a sequence in X that multiplicative converges to a point $y \in X$

(d) for all $c \in E$ with $1 \ll c$, there exists $e \in E$ with $1 \ll e$ such that $q_m(x, z) \ll e$ and $q_m(z, y) \ll e$ imply $m(x, y) \ll c$

Definition B.9 1

A pair (f, g) of self-mappings on a partially ordered set, $(X, \sqsubseteq)$ will be called r-weakly increasing if $f^r x \sqsubseteq g^r f^r x$ and $g^r x \sqsubseteq f^r g^r x$ holds for all $x \in X$, and any $r \in \mathbb{N}$

Definition B.10 1

Let (X, m) be a multiplicative cone b-metric space with co-efficient $s \geq 1$, the function $q_{ms} : X \times X \mapsto E$ will be called a generalized multiplicative c-distance on X if the following conditions are satisfied

(a) $1 \leq q_{ms}(x, y)$ for all $x, y \in X$

(b) $q_{ms}(x, y) \leq [q_{ms}(x, z) \cdot q_{ms}(z, y)]^s$ for all $x, y, z \in X$

(c) for each $x \in X$ and $n \geq 1$, if $q_{ms}(x, y_n) \leq u$ for some $u = u_x \in P$, then, $q_{ms}(x, y) \leq u^s$ whenever $\{y_n\}$ is a sequence in X that multiplicative converges to a point $y \in X$

(d) for all $c \in E$ with $1 \ll c$, there exists $e \in E$ with $1 \ll e$ such that $q_{ms}(x, z) \ll e$ and $q_{ms}(z, y) \ll e$ imply $m(x, y) \ll c$

Example B.11 1

Let (X, m) be a multiplicative cone b-metric space with coefficient $s \geq 1$ and P be a normal multiplicative cone. Put $q_{ms}(x, y) = m(x, y)^{\frac{1}{s}}$ for all $x, y \in X$, then q_{ms} is a generalized multiplicative c-distance

Example B.12 1

Let (X, m) be a multiplicative cone b-metric space with coefficient $s \geq 1$ and P be a normal multiplicative cone. Put $q_{ms}(x, y) = m(u, y)^{\frac{1}{s}}$ for all $x, y \in X$, where u is a fixed point, then q_{ms} is a generalized multiplicative c-distance

2.3 Main Results

Remark B.1 1

Note that Proposition 1.11 [Ampadu, Clement (2016):Higher Order Banach Contraction Principle in Rectangular Multiplicative b-Metric Space, Unpublished] holds if (X, d) is a cone b-metric space. Moreover, if $t(x,y) = \frac{1}{s}d(x,y)$, then $t(x,y)$ is a generalized c-distance, and Proposition 1.11 [Ampadu, Clement (2016):Higher Order Banach Contraction Principle in Rectangular Multiplicative b-Metric Space, Unpublished] still holds, if we take $d(x,y) = st(x,y)$. On the other hand, by Example B.11, with $d_1(x,y) = s\log_a[q_{ms}(x,y)]$, where $a > 1$, we have that (X, d_1) is a cone b-metric space, thus, Proposition 1.11 [Ampadu, Clement (2016):Higher Order Banach Contraction Principle in Rectangular Multiplicative b-Metric Space, Unpublished] holds if $d := d_1$

Theorem B.2 1

Let $(X, \sqsubseteq)$ be a partially ordered set and suppose that (X, m) is a complete multiplicative cone b-metric space. Let q_{ms} be a generalized multiplicative c-distance on X and $f : X \mapsto X$ be a nondecreasing function with respect to $\sqsubseteq$. Suppose that the following assertions hold

(a) there exists $Z \geq 1$ and $q \in [0, \frac{1}{s})$ modified from Proposition 1.11 [Ampadu, Clement (2016):Higher Order Banach Contraction Principle in Rectangular Multiplicative b-Metric Space, Unpublished] such that $q_{ms}(f^r x, f^r y) \leq q_{ms}(x,y)^{Zq^r}$ for all $x, y \in X$ with $x \sqsubseteq y$

(b) there exists $x_0 \in X$ such that $x_0 \sqsubseteq_r fx_0$, that is, $x_0 \sqsubseteq f^r x_0$ for any $r \in \mathbb{N}$

(c) if $\{x_n\}$ is a nondecreasing sequence with respect to $\sqsubseteq$, and converges to x, we have $x_n \sqsubseteq x$ as $n \to \infty$

Then f has a r-fixed point x'. If $v = f^r v$, then $q_{ms}(v,v) = 1$

CHAPTER 2. FIXED POINT THEOREMS IN ORDERED MULTIPLICATIVE CONE B-METRIC SPACE WITH GENERALIZED MULTIPLICATIVE C-DISTANCE

Proof of Theorem B.2 1

If $f^r x_0 = x_0$, then the proof is finished. So we assume that $f^r x_0 \neq x_0$. Since $x_0 \sqsubseteq f^r x_0$ and f is nondecreasing with respect to $\sqsubseteq$, we obtain by induction, $x_0 \sqsubseteq f^r x_0 = x_1 \sqsubseteq \cdots \sqsubseteq f^{rn} x_0 = x_n \sqsubseteq f^{r(n+1)} x_0 = x_{n+1} \sqsubseteq \cdots$. Since $q_{ms}(x_n, x_{n+1}) = q_{ms}(f^r x_{n-1}, f^r x_n) \leq q_{ms}(x_{n-1}, x_n)^{Zq^r}$, we have, $q_{ms}(x_n, x_{n+1}) \leq q_{ms}(x_{n-1}, x_n)^h \leq \cdots \leq q_{ms}(x_0, x_1)^{h^n}$, where $h = Zq^r$, for all $n \geq 1$. Let $k > n$. Then we have,

$$q_{ms}(x_n, x_k) \leq [q_{ms}(x_n, x_{n+1}) \cdot q_{ms}(x_{n+1}, x_k)]^s$$
$$\leq q_{ms}(x_n, x_{n+1})^s \cdot [q_{ms}(x_{n+1}, x_{n+2}) \cdot q_{ms}(x_{n+2}, x_k)]^{s^2}$$
$$\leq q_{ms}(x_n, x_{n+1})^s \cdot q_{ms}(x_{n+1}, x_{n+2})^{s^2} \cdots q_{ms}(x_{k-1}, x_k)^{s^{k-n}}$$
$$\vdots$$
$$\leq q_{ms}(x_0, x_1)^{\frac{sh^n}{1-sh}}$$

where $0 < h = Zq^r < 1$. We show that $\{x_n\}$ is a multiplicative Cauchy sequence in X. Let $c \in E$ with $1 \ll c$ be given, since $\{q_{ms}(x_0, x_1)^{\frac{sh^n}{1-sh}}\}$ converges to 1, it follows from Remark B.7, that there exists a positive integer N such that $q_{ms}(x_0, x_1)^{\frac{sh^n}{1-sh}} \ll c$ for all $n \geq N$. If $e = c$, then, $q_{ms}(x_n, x_{n+1}) \ll e$ and $q_{ms}(x_n, x_k) \ll e$ for any $k > n > N$, thus, $m(x_{n+1}, x_k) \ll c$. Since X is complete, there exists $x' \in X$ such that $x_n \to x'$ as $n \to \infty$. Now notice that $q_{ms}(x_{n-1}, x_k) \leq [q_{ms}(x_{n-1}, x_n) \cdot q_{ms}(x_n, x_k)]^s \leq q_{ms}(x_0, x_1)^{sh^{n-1}} \cdot q_{ms}(x_0, x_1)^{\frac{s^2 h^n}{1-sh}}$, where $h = Zq^r < 1$ for all $k > n > N$. From Definition B.10, it follows that $q_{ms}(x_{n-1}, x') \leq q_{ms}(x_0, x_1)^{s^2 h^{n-1}} \cdot q_{ms}(x_0, x_1)^{\frac{s^3 h^n}{1-sh}}$. On the other hand, $q_{ms}(x_n, f^r x') = q_{ms}(f^r x_{n-1}, f^r x') \leq q_{ms}(x_{n-1}, x')^{Zq^r}$. Since, $\{q_{ms}(x_0, x_1)^{s^2 h^{n-1}} \cdot q_{ms}(x_0, x_1)^{\frac{s^3 h^n}{1-sh}}\}$ converges to 1, it follows from Remark B.7, that there exists a positive integer N_0 such that $q_{ms}(x_{n-1}, x') \ll c$ for all $n \geq N_0$. If $e = c$, then $q_{ms}(x_n, f^r x') \ll e$ and $q_{ms}(x_n, x') \ll e$ for all $n \geq N_0$, and thus $m(f^r x', x') \ll c$, that is, $f^r(x') = x'$. Now if $v = f^r v$, then, $q_{ms}(v, v) = q_{ms}(f^r v, f^r v) \leq q_{ms}(v, v)^{Zq^r}$. Since $1 - Zq^r \neq 0$, it follows that $q_{ms}(v, v) = 1$

Theorem B.3 1

Let $(X, \sqsubseteq)$ be a partially ordered set and suppose that (X, m) is a complete multiplicative cone b-metric space. Let q_{ms} be a generalized multiplicative c-distance on X and $f, g : X \mapsto X$ be r-weakly increasing functions with respect to $\sqsubseteq$. Suppose that the following assertions hold

(a) there exists $Z \geq 1$ and $q \in [0, \frac{1}{s})$ modified from Proposition 1.11 [Ampadu, Clement (2016):Higher Order Banach Contraction Principle in Rectangular Multiplicative b-Metric Space, Unpublished] such that $q_{ms}(f^r x, g^r y) \leq q_{ms}(x, y)^{Zq^r}$ and $q_{ms}(g^r x, f^r y) \leq q_{ms}(x, y)^{Zq^r}$ for all comparable $x, y \in X$

(b) if $\{x_n\}$ is a nondecreasing sequence with respect to $\sqsubseteq$, and converges to x, we have $x_n \sqsubseteq x$ as $n \to \infty$

Then f and g have a common r-fixed point x'. If $v = f^r v = g^r v$, then $q_{ms}(v, v) = 1$

CHAPTER 2. FIXED POINT THEOREMS IN ORDERED MULTIPLICATIVE CONE B-METRIC SPACE WITH GENERALIZED MULTIPLICATIVE C-DISTANCE

Proof of Theorem B.3 1

Let x_0 be an arbitrary point in X and define a sequence $\{x_n\}$ in X as follows: $x_{2n+1} = f^r x_{2n}$ and $x_{2n+2} = g^r x_{2n+1}$ for all $n \geq 0$. Since f and g are r-weakly increasing. We have $x_1 = f^r x_0 \sqsubseteq g^r f^r x_0 = g^r x_1 = x_2$ and $x_2 = g^r x_1 \sqsubseteq f^r g^r x_1 = f^r x_2 = x_3$. Continuing this process, we have, $x_1 \sqsubseteq x_2 \sqsubseteq \cdots \sqsubseteq x_n \sqsubseteq x_{n+1} \sqsubseteq \cdots$. It follows that x_n is non-decreasing or increasing. Now observe that $q_{ms}(x_{2n+1}, x_{2n+2}) = q_{ms}(f^r x_{2n}, g^r x_{2n+1}) \leq q_{ms}(x_{2n}, x_{2n+1})^{Zq^r}$, which implies that $q_{ms}(x_{2n+1}, x_{2n+2}) \leq q_{ms}(x_{2n}, x_{2n+1})^h$, where $h = Zq^r < 1$. By induction, we have, $q_{ms}(x_n, x_{n+1}) \leq q_{ms}(x_{n-1}, x_n)^h \cdots q_{ms}(x_0, x_1)^{h^n}$. Now let $k > n$, as in the proof of the previous theorem, we have, $q_{ms}(x_n, x_k) \leq q_{ms}(x_0, x_1)^{\frac{sh^n}{1-sh}}$. Now we show that $\{x_n\}$ is a multiplicative Cauchy sequence. Let $c \in E$ with $1 \ll c$ be given, since $\{q_{ms}(x_0, x_1)^{\frac{sh^n}{1-sh}}\}$ converges to 1, it follows from Remark B.7, that there exists a positive integer N such that $q_{ms}(x_0, x_1)^{\frac{sh^n}{1-sh}} \ll c$ for all $n \geq N$. If $e = c$, then, $q_{ms}(x_n, x_{n+1}) \ll e$ and $q_{ms}(x_n, x_k) \ll e$ for any $k > n > N$, thus, $m(x_{n+1}, x_k) \ll c$, and $\{x_n\}$ is multiplicative Cauchy. Since X is multiplicative complete, there exists a point $x' \in X$ such that $x_n \to x'$ as $n \to \infty$. Now observe that $q_{ms}(x_{2n+1}, x_k) \leq [q_{ms}(x_{2n+1}, x_{2n+2}) \cdot q_{ms}(x_{2n+2}, x_k)]^s \leq q_{ms}(x_0, x_1)^{sh^{2n+1}} \cdot q_{ms}(x_0, x_1)^{\frac{s^2 h^{2n+2}}{1-sh}}$ for all $k > n > N$, where $0 < h = Zq^r < 1$. From Definition B.10, we have $q_{ms}(x_{2n+1}, x') \leq q_{ms}(x_0, x_1)^{s^2 h^{2n+1}} \cdot q_{ms}(x_0, x_1)^{\frac{s^3 h^{2n+1}}{1-sh}}$. On the other hand, we notice that, $q_{ms}(x_{2n+2}, f^r x') = q_{ms}(g^r x_{2n+1}, f^r x') \leq q_{ms}(x_{2n+1}, x')^{Zq^r}$. Since, $\{q_{ms}(x_0, x_1)^{s^2 h^{2n+1}} \cdot q_{ms}(x_0, x_1)^{\frac{s^3 h^{2n+2}}{1-sh}}\}$ converges to 1, it follows from Remark B.7, that there is a positive integer N such that $q_{ms}(x_0, x_1)^{s^2 h^{2n+1}} \cdot q_{ms}(x_0, x_1)^{\frac{s^3 h^{2n+2}}{1-sh}} \ll c$ for all $n \geq N$. If $e = c$, then, $q_{ms}(x_{2n+1}, x') \ll e$ for any $n > N$. Since, $q_{ms}(x_{2n+1}, f^r x') \leq [q_{ms}(x_{2n+1}, x_{2n+2}) \cdot q_{ms}(x_{2n+2}, f^r x')]^s$, then combining with $q_{ms}(x_{2n+2}, f^r x') = q_{ms}(g^r x_{2n+1}, f^r x') \leq q_{ms}(x_{2n+1}, x')^{Zq^r}$ and Remark B.7, we also deduce that if $e = c$, then $q_{ms}(x_{2n+1}, f^r x') \ll e$ for any $n > N$. Thus, we can conclude that $m(x', f^r x') \ll c$, that is, $f^r x' = x'$. Now notice that $q_{ms}(x_{2n}, x_k) \leq [q_{ms}(x_{2n}, x_{2n+1}) \cdot q_{ms}(x_{2n+1}, x_k)]^s \leq q_{ms}(x_0, x_1)^{sh^{2n}} \cdot q_{ms}(x_0, x_1)^{\frac{s^2 h^{2n+1}}{1-sh}}$, and from Definition B.10, we have, $q_{ms}(x_{2n}, x') \leq q_{ms}(x_0, x_1)^{s^2 h^{2n}} \cdot q_{ms}(x_0, x_1)^{\frac{s^3 h^{2n+1}}{1-sh}}$. Since, $\{q_{ms}(x_0, x_1)^{s^2 h^{2n}} \cdot q_{ms}(x_0, x_1)^{\frac{s^3 h^{2n+1}}{1-sh}}\}$ converges to 1, it follows from Remark B.7, that there is a positive integer N such that $q_{ms}(x_0, x_1)^{s^2 h^{2n}} \cdot q_{ms}(x_0, x_1)^{\frac{s^3 h^{2n+1}}{1-sh}} \ll c$ for all $n \geq N$, thus, if $e = c$, then, $q_{ms}(x_{2n}, x') \ll e$ for all $n > N$. On the other hand, $q_{ms}(x_{2n+1}, g^r x') = q_{ms}(f^r x_{2n}, g^r x') \leq q_{ms}(x_{2n}, x')^{Zq^r}$ and $q_{ms}(x_{2n}, g^r x') \leq [q_{ms}(x_{2n}, x_{2n+1}) \cdot q_{ms}(x_{2n+1}, g^r x')]^s$, thus combining with Remark B.7, if $e = c$, then, $q_{ms}(x_{2n}, g^r x') \ll e$ for all $n > N$. Consequently, $m(g^r x', x') \ll c$, that is, $g^r x' = x'$, and since $f^r x' = x'$, it follows that $g^r x' = x' = f^r x'$, that is, x' is the common r-fixed point of f and g

Example B.4 1

Let $E = [1, \infty)$, and $P = \{x \in E : x \geq 1\}$. Let $X = [0, 1]$ and define a mapping $m : X \times X \mapsto E$ by $m(x, y) = a^{|x-y|^2}$ for all $x, y \in X$ and $a > 1$. Then (X, m) is a multiplicative cone b-metric space. Define a mapping $q_{ms} : X \times X \mapsto E$ by $q_{ms}(x, y) = a^{y^2}$ for all $x, y \in X$ and $a > 1$; let an order relation $\sqsubseteq$ be defined by $x \sqsubseteq y$ iff $x \leq y$, then q_{ms} is a generalized multiplicative c-distance on X. If $f^r x = \frac{x^{2^r}}{4^{2^r}-1}$ for all $x \neq 1$ and $f^r(1) = \frac{1}{2}$, then there exists a modification on Proposition 1.11 [Ampadu, Clement (2016):Higher Order Banach Contraction Principle in Rectangular Multiplicative b-Metric Space, Unpublished] such that f satisfies the assertion of Theorem B.2. Moreover, 0 is the r-fixed point of f

Example B.5 1

In this example, we present an application of Theorem B.2 - showing existence of solution of an integral equation. Let $X = C(I, \mathbb{R}^n)$, $E = \mathbb{R}^n$, $P = \{(x_1, x_2, \cdots, x_n) : x_i \geq 1, i = 1, 2, \cdots, n\}$, and define $m : X \times X \mapsto E$ by $m(x, y) = \{m(x, y)_i\}_{i=1}^n$, where $m(x, y)_i = a^{\sup_{t \in I} |x(t) - y(t)|^2}$, $i = 1, 2, \cdots, n$, for every $x, y \in X$, and $a > 1$. Then (X, m) is a multiplicative cone b-metric space with coefficient $s = 2$. Define a mapping $q_{ms} : X \times X \mapsto E$ by $q_{ms}(x, y) = \{q_{ms}(x, y)_i\}_{i=1}^n$, where $q_{ms}(x, y)_i = a^{\sup_{t \in I} |y(t)|^2}$, $i = 1, 2, \cdots, n$, for every $x, y \in X$, and $a > 1$, and let an order relation $\sqsubseteq$ be given by $x \sqsubseteq y$ iff $a^{\sup_{t \in I} |x(t)|} \leq a^{\sup_{t \in I} |y(t)|}$, where $a > 1$. Then q is a generalized multiplicative c-distance on X. Now let I be the closed unit interval $[0, 1]$ in $\mathbb{R}$, and consider the following integral equation, $x(t) = \int_0^t g(s, x(s))ds$, $t \in I$, where $g : I \times \mathbb{R}^n \mapsto \mathbb{R}^n$ is such that $g(s, \cdot)$ is increasing for every $s \in I$. Let there exists a modification on Proposition 1.11 [Ampadu, Clement (2016):Higher Order Banach Contraction Principle in Rectangular Multiplicative b-Metric Space, Unpublished] such that $\{(|g(s, y)_i|)\}_{i=1}^n \leq \sqrt{Zq^r}\{|y(s)|, \cdots, |y(s)|\}$ for every $s \in I$ and $x, y \in X$, and any $r \in \mathbb{N}$, then the integral equation, $x(t) = \int_0^t g(s, x(s))ds$, $t \in I$, has a solution in $C(I, \mathbb{R}^n)$. To see this, define $T^r x(t) = \int_0^t g(s, x(s))ds$ for any $r \in \mathbb{N}$. Now for each $x, y \in X$, we have,

$$q_{ms}(T^r x, T^r y) = (a^{\sup_{t \in I} |[T^r y](t)|^2}, \cdots, a^{\sup_{t \in I} |[T^r y](t)|^2})$$
$$\leq (a^{\sup_{t \in I} (\int_0^t |g(s,y(s))|)^2}, \cdots, a^{\sup_{t \in I} (\int_0^t |g(s,y(s))|)^2})$$
$$\leq (a^{\sup_{t \in I} (\int_0^t \sqrt{Zq^r} |y(s)|)^2}, \cdots, a^{\sup_{t \in I} (\int_0^t \sqrt{Zq^r} |y(s)|)^2})$$
$$\leq (a^{Zq^r \sup_{t \in I} (\int_0^t |y(s)|^2)}, \cdots, a^{Zq^r \sup_{t \in I} (\int_0^t |y(s)|^2)})$$
$$\leq (a^{Zq^r \sup_{t \in I} (\int_0^t \sup |y(s)|^2)}, \cdots, a^{Zq^r \sup_{t \in I} (\int_0^t \sup |y(s)|^2)})$$
$$\leq a^{Zq^r (\sup |y(s)|^2, \cdots, \sup |y(s)|^2) \sup_{t \in I} \int_0^t 1 ds}$$
$$\leq q_{ms}(x, y)^{Zq^r}$$

Thus according to Theorem B.2, the integral equation, $x(t) = \int_0^t g(s, x(s))ds$, $t \in I$, has a solution

2.4 Exercises

Exercise B.1 1

Consider Remark B.1, prove that (X, d) is a cone b-metric space with coefficient $s \geq 1$, where $d(x, y) = d_1(x, y) = s \log_a(q_{ms}(x, y))$ for every $x, y \in X$ and $a > 1$ and $q_{ms}(x, y)$ is a generalized multiplicative c-distance on X

Exercise B.2 1

Verify Example B.4 by taking the following steps

(a) Prove that (X, m) is a multiplicative cone b-metric space

(b) Prove that q_{ms} is a generalized multiplicative c-distance on X

(c) Prove that f satisfies the assertion of Theorem B.2 by considering the following cases

    (i) $x = y = 1$

    (ii) $x \neq 1$ and $y = 1$

    (iii) $x \neq 1$ and $y \neq 1$

CHAPTER 2. FIXED POINT THEOREMS IN ORDERED MULTIPLICATIVE CONE B-METRIC SPACE WITH GENERALIZED MULTIPLICATIVE C-DISTANCE

Exercise B.3 1

Let the setting be this chapter. Deduce the higher-order version of Theorem 2.11 [Baoguo Bao et.al, Fixed point theorems on generalized c-distance in ordered cone b-metric spaces, Int. J. Nonlinear Anal. Appl. 6 (2015) No. 1, 9-22]

Exercise B.4 1

Let the setting be this chapter. Deduce the higher-order version of Theorem 4.1 [Zaid Mohammed Fadail et.al, Generalized c-Distance in Cone b-Metric Spaces and Common Fixed Point Results for Weakly Compatible Self-Mappings, International Journal of Mathematical Analysis Vol. 9, 2015, no. 32, 1593 - 1607]

Exercise B.5 1

As a consequence of the previous exercise. Deduce the higher-order version of the following

(a) Corollary 4.4 [Zaid Mohammed Fadail et.al, Generalized c-Distance in Cone b-Metric Spaces and Common Fixed Point Results for Weakly Compatible Self-Mappings, International Journal of Mathematical Analysis Vol. 9, 2015, no. 32, 1593 - 1607]

(b) Corollary 4.5 [Zaid Mohammed Fadail et.al, Generalized c-Distance in Cone b-Metric Spaces and Common Fixed Point Results for Weakly Compatible Self-Mappings, International Journal of Mathematical Analysis Vol. 9, 2015, no. 32, 1593 - 1607]

(c) Corollary 4.6 [Zaid Mohammed Fadail et.al, Generalized c-Distance in Cone b-Metric Spaces and Common Fixed Point Results for Weakly Compatible Self-Mappings, International Journal of Mathematical Analysis Vol. 9, 2015, no. 32, 1593 - 1607]

Exercise B.6 1

Deduce the higher-order version of Theorem 4.7 [Zaid Mohammed Fadail et.al, Generalized c-Distance in Cone b-Metric Spaces and Common Fixed Point Results for Weakly Compatible Self-Mappings, International Journal of Mathematical Analysis Vol. 9, 2015, no. 32, 1593 - 1607], and show it is a consequence of Exercise B.4

Exercise B.7 1

Give an example to support the Theorem obtained from Exercise B.4

> **Exercise B.8 1**
>
> Taking inspiration from [Sushanta Kumar Mohanta, SOME FIXED POINT THEOREMS VIA GENERALIZED c-DISTANCE IN ORDERED CONE METRIC SPACES, Adv. Fixed Point Theory, 4 (2014), No. 1, 102-116] we introduce the following: Let $s = 1$ in Definition B.10, that is, let (X, m) be a multiplicative cone metric space. A function $q_m : X \times X \mapsto E$ will be called a generalized multiplicative c-distance of order j on X, where $j \in \mathbb{N}$ if in addition to satisfying (a), (c), and (d) of Definition B.10, it satisfies $q_m(x, z) \leq \prod_{i=0}^{j} q_m(x_i, x_{i+1})$ for all $x, z \in X$ and for all distinct points $x_i \in X$, $i \in \{1, 2, 3, \cdots, j\}$ each of them different from $x := x_0$ and $z := x_{j+1}$. Let the setting be ordered multiplicative cone metric space. Using the definition of generalized multiplicative c-distance of order j, deduce the higher-order version of the following
>
> (a) Theorem 3.6 [Sushanta Kumar Mohanta, SOME FIXED POINT THEOREMS VIA GENERALIZED c-DISTANCE IN ORDERED CONE METRIC SPACES, Adv. Fixed Point Theory, 4 (2014), No. 1, 102-116]
>
> (b) Theorem 3.7 [Sushanta Kumar Mohanta, SOME FIXED POINT THEOREMS VIA GENERALIZED c-DISTANCE IN ORDERED CONE METRIC SPACES, Adv. Fixed Point Theory, 4 (2014), No. 1, 102-116]
>
> (c) Theorem 3.8 [Sushanta Kumar Mohanta, SOME FIXED POINT THEOREMS VIA GENERALIZED c-DISTANCE IN ORDERED CONE METRIC SPACES, Adv. Fixed Point Theory, 4 (2014), No. 1, 102-116]
>
> (d) Theorem 3.9 [Sushanta Kumar Mohanta, SOME FIXED POINT THEOREMS VIA GENERALIZED c-DISTANCE IN ORDERED CONE METRIC SPACES, Adv. Fixed Point Theory, 4 (2014), No. 1, 102-116]

2.5 References

(1) Ampadu, Clement (2015): Multiplicative Soft Cone Metric Spaces and Some Fixed Point Theorems for Multiplicative Expanding Mappings, Unpublished

(2) Ampadu, Clement (2015): Multiplicative Soft Cone Metric Spaces and Some Fixed Point Theorems for Multiplicative Contraction Mappings, Unpublished

(3) Ampadu, Clement (2016): Higher Order Banach Contraction Principle in Rectangular Multiplicative b-Metric Space, Unpublished

(4) Baoguo Bao et.al, Fixed point theorems on generalized c-distance in ordered cone b-metric spaces, Int. J. Nonlinear Anal. Appl. 6 (2015) No. 1, 9-22

(5) Zaid Mohammed Fadail et.al, Generalized c-Distance in Cone b-Metric Spaces and Common Fixed Point Results for Weakly Compatible Self-Mappings, International Journal of Mathematical Analysis Vol. 9, 2015, no. 32, 1593 - 1607

(6) Sushanta Kumar Mohanta, SOME FIXED POINT THEOREMS VIA GENERALIZED c-DISTANCE IN ORDERED CONE METRIC SPACES, Adv. Fixed Point Theory, 4 (2014), No. 1, 102-116

Chapter 3

r-Points of Coincidence and Common r-Fixed Points for Higher-Order Expansive Type Multiplicative Contractions in Multiplicative Cone b-Metric Space

3.1 Brief Summary

> **Abstract C.1 1**
>
> In this chapter we obtain some sufficient conditions for existence of r-points of coincidence and common r-fixed points for a pair of self-mappings satisfying some higher-order expansive type conditions in multiplicative cone b-metric space

3.2 Preliminaries

Multiplicative cone in soft metric spaces was given in [Ampadu,Clement (2015): Multiplicative Soft Cone Metric Spaces and Some Fixed Point Theorems for Multiplicative Expanding Mappings, Unpublished; Ampadu, Clement (2015): Multiplicative Soft Cone Metric Spaces and Some Fixed Point Theorems for Multiplicative Contraction Mappings, Unpublished]. It follows from these papers that we have the following

> **Definition C.1 1**
>
> Let E be a real Banach space and P be a subset of E. We will say P is a *multiplicative cone* if
>
> (a) P is closed, nonempty, and satisfies $P \neq \{1\}$
>
> (b) $x^a \cdot y^b \in P$ for all $x, y \in X$ and non-negative real numbers a, b
>
> (c) $x \in P$ and $\frac{1}{x} \in P$ imply $x = 1$, that is, $P \cap \frac{1}{P} = 1$

Definition C.2 1

Given a multiplicative cone $P \subset E$, we define a partial ordering $\preceq$ with respect to P by $x \preceq y$ iff $\frac{y}{x} \in P$. We will write $x < y$ if $x \preceq y$ and $x \neq y$, and $x \ll y$ if $\frac{y}{x} \in int\ P$, where $int\ P$ denotes the interior of P. A multiplicative cone P will be multiplicative normal if there is a number $N > 0$ such that for all $x, y \in P$, $1 \leq x \leq y$ implies $\|x\| \leq \|y\|^N$. The least positive number satisfying the inequality, $\|x\| \leq \|y\|^N$, will be called the multiplicative normal constant of P

Definition C.3 1

Let X be a nonempty set. Suppose that the mapping $m : X \times X \mapsto E$ satisfies

(a) $1 \preceq m(x, y)$ for all $x, y \in X$ and $m(x, y) = 1$ iff $x = y$

(b) $m(x, y) = m(y, x)$ for all $x, y \in X$

(c) $m(x, y) \preceq m(x, z) \cdot m(z, y)$ for all $x, y, z \in X$

Then m will be called a multiplicative cone metric on X and (X, m) will be called a multiplicative cone metric space

Definition C.4 1

Let X be a nonempty set and $s \geq 1$ be a real number. Suppose that the mapping $m : X \times X \mapsto E$ satisfies

(a) $1 \preceq m(x, y)$ for all $x, y \in X$ and $m(x, y) = 1$ iff $x = y$

(b) $m(x, y) = m(y, x)$ for all $x, y \in X$

(c) $m(x, y) \preceq [m(x, z) \cdot m(z, y)]^s$ for all $x, y, z \in X$

Then m will be called multiplicative cone b-metric on X and (X, m) will be called a multiplicative cone b-metric space

Example C.5 1

Let $X = \{-1, 0, 1\}$, $E = \mathbb{R}^2$, and $P = \{(x, y) : x \geq 1, y \geq 1\}$. Define $m : X \times X \mapsto E$ by $m(x, y) = m(y, x)$ for all $x, y \in X$, $m(x, x) = 1$ for all $x \in X$, and $m(-1, 0) = (a^3, a^3)$ for some $a > 1$; $m(-1, 1) = m(0, 1) = (a, a)$ for some $a > 1$. Then (X, m) is a multiplicative cone b-metric space, but not a multiplicative cone metric space since the multiplicative triangle inequality does not hold. Indeed, we have, $m(-1, 1) \cdot m(1, 0) = (a, a) \cdot (a, a) = (a^2, a^2) \preceq (a^3, a^3) = m(-1, 0)$. Note that (X, m) is a multiplicative cone b-metric space with coefficient $s = \frac{3}{2}$

Example C.6 1

Let $E = \mathbb{R}^2$, $P = \{(x, y) : x \geq 1, y \geq 1\} \subseteq E$, $X = \mathbb{R}$ and $m : X \times X \mapsto E$ such that $m(x, y) = (a^{|x-y|^p}, a^{\alpha|x-y|^p})$, where $\alpha \geq 0$; $a, p > 1$ are three constants. Then (X, m) is a multiplicative cone b-metric space with co-efficient $s = 2^{p-1}$, but not a multiplicative cone metric space.

Definition C.7 1

Let (X, m) be a cone b-metric space. The sequence $\{x_n\}$ will be called

(a) a multiplicative convergent sequence if for every $c \in E$ with $1 \ll c$, there is $n_0 \in \mathbb{N}$ such that for all $n \geq n_0$, $m(x_n, x) \ll c$ for some $x \in X$

(b) a multiplicative Cauchy sequence if for all $c \in E$ with $1 \ll c$, there is $n_0 \in \mathbb{N}$ such that for all $n, k \geq n_0$, $m(x_n, x_k) \ll c$

Definition C.8 1

The cone b-metric space (X, m) will be called multiplicative complete if every multiplicative Cauchy sequence in X is multiplicative convergent in X

Remark C.9 1

Let (X, m) be a multiplicative cone b-metric space over the ordered real Banach space E with multiplicative cone P. In the sequel some of the following properties will be useful

(a) If $a \preceq b$ and $b \ll c$, then $a \ll c$

(b) If $a \ll b$ and $b \ll c$, then $a \ll c$

(c) If $1 \preceq u \ll c$ for each $c \in int(P)$, then $u = 1$

(d) If $c \in int(P)$, $1 \preceq a_n$ and $a_n \to 1$, then there exist n_0 such that for all $n > n_0$ we have $a_n \ll c$

(e) Let $1 \ll c$. If $1 \preceq m(x_n, x) \preceq b_n$ and $b_n \to 1$, then eventually $m(x_n, x) \ll c$, where $\{x_n\}$ and x are a sequence and a given point in X

(f) If $1 \preceq a_n \preceq b_n$ and $a_n \to a$ and $b_n \to b$, then $a \preceq b$ for each multiplicative cone P

(g) If E is a real Banach space with multiplicative cone P and if $a \preceq a^\lambda$, where $a \in P$ and $0 \leq \lambda < 1$, then $a = 1$

(h) $int(P)^\alpha \subseteq int(P)$ for all $\alpha > 0$

(i) For each $\delta > 1$ and $x \in int(P)$ there is $0 < \gamma < 1$ such that $\|x^\gamma\| < \delta$

(j) For each $1 \ll c_1$ and $c_2 \in P$, there is an element $1 \ll d$ such that $c_1 \ll d$ and $c_2 \ll d$

(k) For each $1 \ll c_1$ and $1 \ll c_2$, there is an element $1 \ll e$ such that $e \ll c_1$ and $e \ll c_2$

Definition C.10 1

Let (X, m) be a multiplicative cone b-metric space and let $T : X \mapsto X$ be a given mapping. We say T is r-continuous at $x_0 \in X$ if $T^r x_n \to T^r x_0$ as $n \to \infty$ for every sequence $\{x_n\}$ in X satisfying $x_n \to x_0$ as $n \to \infty$. If T is r-continuous at each point $x_0 \in X$, then we say T is r-continuous on X

Definition C.11 1

Let (X, m) be a multiplicative cone b-metric space with constant $s \geq 1$. A map $T : X \mapsto X$ will be called higher-order expansive if it satisfies $m(T^r x, T^r y) \succeq m(x, y)^{Zq^r}$, where Z is given by Proposition 1.11[Ampadu, Clement (2016): Higher Order Banach Contraction Principle in Rectangular Multiplicative b-Metric Space, Unpublished], and however $q > s$

Definition C.12 1

Let T and S be self-mappings of a set X. If $y = T^r x = S^r x$ for some $x \in X$ and any $r \in \mathbb{N}$, then we say that x is a r-coincidence point of T and S and y is called a r-point of coincidence of S and T

Definition C.13 1

The mappings $T, S : X \mapsto X$ are r-weakly compatible for any $r \in \mathbb{N}$, if for every $x \in X$, the following holds, $T^r(S^r x) = S^r(T^r x)$ whenever $S^r x = T^r x$

Proposition C.14 1

Let S and T be r-weakly compatible selfmaps of a nonempty set X. If S and T have a unique r-point of coincidence $y = S^r x = T^r x$, then y is the unique common r-fixed point of S and T.

3.3 Main Results

Theorem C.1 1

Let (X, m) be a multiplicative cone b-metric space with coefficient $s \geq 1$. Suppose the mappings $f, g : X \mapsto X$ satisfy $g^r(X) \subseteq f^r(X)$ for any $r \in \mathbb{N}$, either $f^r(X)$ or $g^r(X)$ is complete for any $r \in \mathbb{N}$. Further, suppose that $m(f^r x, f^r y) \succeq m(g^r x, g^r y)^{Z^\star q^r}$, for all $x, y \in X$ where $Z^\star$ is a certain modificiation on Proposition 1.11[Ampadu, Clement (2016): Higher Order Banach Contraction Principle in Rectangular Multiplicative b-Metric Space, Unpublished], and however $q > s$. Then f and g have a r-point of coincidence in X. Moreover, if $Z^\star q^r > 1$, then the r-point of coincidence is unique. If f and g are r-weakly compatible and $Z^\star q^r > 1$, then f and g have a unique common r-fixed point in X

Proof of Theorem C.1 1

Let $x_0 \in X$ and choose $x_1 \in X$ such that $g^r x_0 = f^r x_1$. This is possible since $g^r(X) \subseteq f^r(X)$. Continuing this process, we can construct a sequence $\{x_n\}$ in X such that $f^r x_n = g^r x_{n-1}$ for all $n \geq 1$. Now notice that $m(g^r x_{n-1}, g^r x_n) = m(f^r x_n, f^r x_{n+1}) \succeq m(g^r x_n, g^r x_{n+1})^{Z^\star q^r}$. It follows that, $m(g^r x_n, g^r x_{n+1}) \preceq m(g^r x_{n-1}, g^r x_n)^\lambda$, where $\lambda = \frac{1}{Z^\star q^r}$. Note that $s\lambda < 1$. By induction we obtain $m(g^r x_n, g^r x_{n+1}) \preceq m(g^r x_0, g^r x_1)^{\lambda^n}$ for all $n \geq 1$. Now let $k, n \in \mathbb{N}$ with $k > n$. Now notice that

$$m(g^r x_n, g^r x_k) \preceq [m(g^r x_n, g^r x_{n+1}) \cdot m(g^r x_{n+1}, g^r x_k)]^s$$
$$\preceq m(g^r x_n, g^r x_{n+1})^s \cdot \cdots \cdot [m(g^r x_{k-2}, g^r x_{k-1}) \cdot m(g^r x_{k-1}, g^r x_k)]^{s^{k-n-1}}$$
$$\preceq m(g^r x_0, g^r x_1)^{[s\lambda^n + \cdots + s^{k-n-1}\lambda^{k-1}]}$$
$$= m(g^r x_0, g^r x_1)^{s\lambda^n[1 + s\lambda + (s\lambda)^2 + \cdots + (s\lambda)^{k-n-1}]}$$
$$\preceq m(g^r x_0, g^r x_1)^{\frac{s\lambda^n}{1-s\lambda}}$$

Note that $m(g^r x_0, g^r x_1)^{\frac{s\lambda^n}{1-s\lambda}} \to 1$ as $n \to \infty$. Let $1 \ll c$ be given. Then we can find $t_0 \in \mathbb{N}$ such that $m(g^r x_0, g^r x_1)^{\frac{s\lambda^n}{1-s\lambda}} \ll c$ for each $n > t_0$, it follows that $m(g^r x_n, g^r x_k) \preceq m(g^r x_0, g^r x_1)^{\frac{s\lambda^n}{1-s\lambda}} \ll c$ for all $k > n > t_0$. So the sequence $\{g^r x_n\}$ is multiplicative Cauchy in $g^r(X)$. Suppose that $g^r(X)$ is a complete subspace of X, then there exists $y \in g^r(X) \subseteq f^r(X)$ such that $g^r x_n \to y$ and also $f^r x_n \to y$. In the case $f^r(X)$ is complete, this holds with $y \in f^r(X)$. Let $u \in X$ be such that $f^r u = y$. For $1 \ll c$, one can choose a natural number $v_0 \in \mathbb{N}$ such that $m(y, g^r x_n) \ll c^{\frac{1}{2s}}$ and $m(f^r x_n, f^r u) \ll c^{\frac{Z^\star q^r}{2s}}$ for all $n > v_0$. Now observe that

$$m(g^r x_{n-1}, f^r u) = m(f^r x_n, f^r u) \succeq m(g^r x_n, g^r u)^{Z^\star q^r}$$

If $Z^\star q^r \neq 0$, it follows that $m(g^r x_n, g^r u) \preceq m(g^r x_{n-1}, f^r u)^{\frac{1}{Z^\star q^r}}$. Thus for all $n > v_0$, we have,

$$m(y, g^r u) \preceq [m(y, g^r x_n) \cdot m(g^r x_n, g^r u)]^s$$
$$\preceq m(y, g^r x_n)^s \cdot m(g^r x_{n-1}, f^r u)^{\frac{s}{Z^\star q^r}}$$
$$= m(y, g^r x_n)^s \cdot m(f^r x_n, f^r u)^{\frac{s}{Z^\star q^r}}$$
$$\ll c$$

It follows that $m(y, g^r u) = 1$, that is, $g^r u = y$, and hence $f^r u = g^r u = y$. Therefore y is a r-point of coincidence of f and g. Now we suppose that $Z^\star q^r > 1$. Let v be another r-point of coincidence of f and g. So, $f^r x = g^r x = v$ for some $x \in X$, then, $m(y, v) = m(f^r u, f^r x) \succeq m(g^r u, g^r x)^{Z^\star q^r} = m(y, v)^{Z^\star q^r}$ which implies that $m(y, v) \preceq m(y, v)^{\frac{1}{Z^\star q^r}}$. By Remark C.9(g), we have $m(y, v) = 1$, that is, $y = v$. Therefore, f and g have a unique r-point of coincidence in X. If f and g are r-weakly compatible, then by Proposition C.14, f and g have a unique common r-fixed point in X.

If f is the identity in the previous theorem, then we get the following

Corollary C.2 1

Let (X, m) be a multiplicative cone b-metric space with coefficient $s \geq 1$. Suppose the mappings $g : X \mapsto X$ satisfy $m(g^r x, g^r y) \preceq m(x, y)^{Zq^r}$, for all $x, y \in X$ where $Z \geq 1$ is given by Proposition 1.11[Ampadu, Clement (2016): Higher Order Banach Contraction Principle in Rectangular Multiplicative b-Metric Space, Unpublished], where $q \in [0, \frac{1}{s})$. Then g has a unique r-fixed point in X. Furthermore, the iterative sequence $\{g^{rn}x\}$ converges to the r-fixed point

If g is the identity in the previous theorem, then we get the following

Corollary C.3 1

Let (X, m) be a multiplicative cone b-metric space with coefficient $s \geq 1$. Suppose the mappings $f : X \mapsto X$ is r-onto, that is, f^r is onto for any $r \in \mathbb{N}$, and satisfy $m(f^r x, f^r y) \succeq m(x,y)^{Zq^r}$, for all $x, y \in X$ where Z is given by Proposition 1.11[Ampadu, Clement (2016): Higher Order Banach Contraction Principle in Rectangular Multiplicative b-Metric Space, Unpublished], and however $q > s$. Then f has a unique r-fixed point

3.4 Exercises

Exercise C.1 1

Verify (X, m) is a multiplicative cone b-metric space with coefficient $s = \frac{3}{2}$ in Example C.5

Exercise C.2 1

Verify (X, m) is a multiplicative cone b-metric space with coefficient $s = 2^{p-1}$, but not a multiplicative cone metric space in Example C.6

Exercise C.3 1

Let the setting be this chapter. Obtain the higher-order version of Corollary 2.6 [Sushanta Kumar Mohanta and Rima Maitra, Coincidence Points And Common Fixed Points For Expansive Type Mappings In Cone b-Metric Spaces, Applied Mathematics E-Notes, 14(2014), 200-208]

Exercise C.4 1

Let the setting be complete multiplicative b-metric space. Obtain the higher-order version of Corollary 3.6 [Sushanta Kumar Mohanta, Coincidence Points and Common Fixed Points for Expansive Type Mappings in b-Metric Spaces, Iranian Journal of Mathematical Sciences and Informatics Vol. 11, No. 1 (2016), pp 101-113]

Exercise C.5 1

Deduce that Theorem C.1 is a special case of Exercise C.3

3.5 References

(1) Ampadu,Clement (2015): Multiplicative Soft Cone Metric Spaces and Some Fixed Point Theorems for Multiplicative Expanding Mappings, Unpublished

(2) Ampadu, Clement (2015): Multiplicative Soft Cone Metric Spaces and Some Fixed Point Theorems for Multiplicative Contraction Mappings, Unpublished

(3) Ampadu, Clement (2016): Higher Order Banach Contraction Principle in Rectangular Multiplicative b-Metric Space, Unpublished

(4) Sushanta Kumar Mohanta and Rima Maitra, Coincidence Points And Common Fixed Points For Expansive Type Mappings In Cone b-Metric Spaces, Applied Mathematics E-Notes, 14(2014), 200-208

(5) Sushanta Kumar Mohanta, Coincidence Points and Common Fixed Points for Expansive Type Mappings in b-Metric Spaces, Iranian Journal of Mathematical Sciences and Informatics Vol. 11, No. 1 (2016), pp 101-113

Chapter 4

Common r-Fixed Point Theorems in b-TVS Multiplicative Cone Metric Space

4.1 Brief Summary

> **Abstract D.1 1**
>
> In this chapter we introduce a notion of b-cone multiplicative metric space over a topological vector space and use it to extend some results of Abbas and Jungck [M. Abbas, G. Jungck, Common fixed point results for non-commuting mappings without continuity in cone metric spaces, J. Math. Anal. Appl., 341, 2008] to this setting.

4.2 Preliminaries

> **Notation D.1 1**
>
> E will denote a topological vector space (for short TVS) with one vector 1_E

> **Definition D.2 1**
>
> A subset K of E will be called a multiplicative cone if:
>
> (a) K is closed, nonempty and $K \neq \{1_E\}$
>
> (b) $a, b \in \mathbb{R}$, $a, b \geq 0$ and $x, y \in K$ imply $x^a \cdot y^b \in K$
>
> (c) $K \cap \frac{1}{K} = \{1_E\}$

> **Definition D.3 1**
>
> For a multiplicative cone $K \subseteq E$ we define a partial ordering $\leq_K$ with respect to K by $x \leq_K y$ iff $\frac{y}{x} \in K$. We shall write $x <_K y$ to indicate $x \leq_K y$ but $x \neq y$, while $x \ll y$ will stand for $\frac{y}{x} \in int(K)$, where $int(K)$ denotes the interior of K

> **Remark D.4 1**
>
> In the following, unless otherwise specified, we always suppose that Y is a locally multiplicative convex Hausdorff TVS with one vector, K a multiplicative cone in Y with $int(K) \neq \emptyset$, $e \in int(K)$ and $\leq_K$ a partial ordering with respect to K

The nonlinear scalarization function has been studied by many authors, see for example [Chen,G.Y., Huang,X.X., Yang, X.Q.: Vector Optimization, Springer-Verlag,Berlin, Heidelberg, Germany, 2005; Du Wei-Shih: On some nonlinear problems induced by an abstract maximal element principle, J. Math. Anal. Appl.,347 (2008), 391-399; Du Wei-Shih, A note on cone metric fixed point theory and its equivalence, Nonlinear Analysis, 72 (5)(2010),2259-2261], and in the present chapter we introduce a nonlinear scalarization type function useful for our purposes.

Definition D.5 1

A function $\zeta_e : Y \mapsto \mathbb{R}$ will be called a nonlinear scalarization type function if it satisfies $\zeta_e(y) = \inf\{r \in \mathbb{R} | y \in \frac{e^r}{K}\}$

Taking inspiration from [G. Y. Chen, X. X. Huang, X. Q. Yang, Vector Optimization, Springer-Verlag, Berlin, Heidelberg, Germany, 2005] we have the following

Lemma D.6 1

For each $r \in \mathbb{R}$ and $y \in Y$, the following statements are satisfied

(a) $\zeta_e(y) \leq r$ iff $y \in \frac{e^r}{K}$

(b) $\zeta_e(y) > r$ iff $y \notin \frac{e^r}{K}$

(c) $\zeta_e(y) \geq r$ iff $y \notin \frac{e^r}{int(K)}$

(d) $\zeta_e(y) < r$ iff $y \in \frac{e^r}{int(K)}$

(e) $\zeta_e(\cdot)$ is positively homogeneous and continuous on Y

(f) if $y_1 \in y_2 \cdot K$, then $\zeta_e(y_2) \leq \zeta_e(y_1)$

(g) $\zeta_e(y_1 \cdot y_2) \leq \zeta_e(y_1) \cdot \zeta_e(y_2)$ for all $y_1, y_2 \in Y$

Definition D.7 1

Let X be a nonempty set. Suppose the mapping $m : X \times X \mapsto Y$ satisfies:

(a) $1 \leq_K m(x,y)$ for all $x, y \in X$ and $m(x,y) = 1$ iff $x = y$

(b) $m(x,y) = m(y,x)$ for all $x, y \in X$

(c) there exists $a \geq 1$ such that $m(x,y) \leq_K [m(x,z) \cdot m(z,y)]^a$ for all $x, y, z \in X$

Then m will be called a b-TVS multiplicative cone metric on X and (X, m) will be called b-TVS multiplicative cone metric space

Example D.8 1

Any TVS multiplicative cone metric space is a b-TVS multiplicative cone metric space.

Example D.9 1

Let $L_p(0 < p < 1)$ be the space of all real functions $x(t)$, $t \in [0,1]$, such that $\int_0^1 |x(t)|^p dt < \infty$, $P = \{(x,y) \in E | x, y \geq 1\} \subset \mathbb{R}^2$, and $\alpha \geq 0$. Then $m : L_p \times L_p \mapsto \mathbb{R}^2$ defined by $m(x,y) = (v^{\int_0^1 |x(t)-y(t)|^p dt^{\frac{1}{p}}}, v^{\alpha \int_0^1 |x(t)-y(t)|^p dt^{\frac{1}{p}}})$ for some $v > 1$ is a b-TVS multiplicative cone metric space on L_p with $a = 2^p$

CHAPTER 4. COMMON R-FIXED POINT THEOREMS IN B-TVS MULTIPLICATIVE CONE METRIC SPACE

Definition D.10 1

Let $\{x_n\}$ be a sequence in X and $x \in X$

(a) We will say x_n b-TVS cone multiplicative converges to x whenever for every $1 \ll c \in E$ there exists $N \in \mathbb{N}$ such that $m(x_n, x) \ll c$, for all $n > N$

(b) We will say x_n is a b-TVS cone multiplicative Cauchy sequence whenever for every $1 \ll c \in E$ there exists $N \in \mathbb{N}$ such that $m(x_n, x_m) \ll c$, for all $n, m > N$

(c) (X, m) is b-TVS multiplicative complete if every b-TVS cone multiplicative Cauchy sequence is b-TVS cone multiplicative convergent.

Definition D.11 1

Let f and g be two self-maps of a set X. If $w = f^r x = g^r x$ for some $x \in X$ and any $r \in \mathbb{N}$, then we will say that x is a r-coincidence point of f and g, and w will be called a r-point of coincidence of f and g

Taking inspiration from [M. Abbas, G. Jungck, Common fixed point results for noncommuting mappings without continuity in cone metric spaces, J. Math. Anal. Appl., 341, 2008] we have the following

Proposition D.12 1

Let f and g be r-weakly compatible selfmaps of a set X. If f and g have a unique r-point of coincidence $w = f^r x = g^r x$ for any $r \in \mathbb{N}$, then w is the unique common r-fixed point of f and g.

4.3 Main Results

Theorem D.1 1

Let (X, m) be a b-TVS multiplicative cone metric space. Then $M : X \times X \mapsto [1, \infty)$ defined by $M = \zeta_e \circ m$ is a multiplicative b-metric

Proof of Theorem D.1 1

Since $1 \leq_K m(x, y)$, we have $m(x, y) \notin \frac{1}{int(k)}$ for all $x, y \in X$. By Lemma 4.6(c) it follows that $M(x, y) \geq 1$ for all $x, y \in X$. If $M(x, y) = 1$, then, from Lemma D.6(a) we have that $m(x, y) \in \frac{1}{K} \cap K = \{1\}$, that is, $x = y$. Conversely, if $x = y$, then, $m(x, y) = 1$. Hence, $M(x, y) = \zeta_e(1) = 1$. It is clear that $M(x, y) = M(y, x)$. Since $m(x, y) \leq_K [m(x, y) \cdot m(y, z)]^a$, we have that, $[m(x, y) \cdot m(y, z)]^a \in K \cdot m(x, y)$. Then via Lemma D.6 (f,g,h), we obtain that $\zeta_e(m(x, y)) \leq \zeta_e([m(x, z) \cdot m(z, y)]^a) \leq [\zeta_e(m(x, z)) \cdot \zeta_e(m(z, y))]^a$ for all $x, y \in X$. Thus, $M(x, y) \leq M(x, z) \cdot M(z, y)$ for all $x, y \in X$.

Theorem D.2 1

Let (X, m) be a b-TVS multiplicative cone metric space and $x \in X$ and $\{x_n\}_{n \in \mathbb{N}}$ a sequence in X. Then the following holds:

(a) if $\{x_n\}$ b-TVS cone multiplicative converges to x, then $M(x_n, x) \to 1$ as $n \to \infty$

(b) if $\{x_n\}$ is a b-TVS cone multiplicative Cauchy sequence, then $\{x_n\}$ is a multiplicative Cauchy sequence with respect to the multiplicative b-metric M

(c) if (X, m) is b-TVS cone multiplicative complete, then (X, m) is a complete multiplicative b-metric space

Proof of Theorem D.2 1

(a) Let $\epsilon > 1$ be given. Since $\{x_n\}$ b-TVS cone multiplicative converges to x, it follows that there exists $n_0 \in \mathbb{N}$ such that $m(x_n, x) \ll e^\epsilon$ for all $n \geq n_0$. Therefore, $\frac{e^\epsilon}{m(x_n,x)} \in int(K)$. Thus, $m(x_n, x) \in \frac{e^\epsilon}{int(K)}$. By Lemma D.6(d), we have, $M(x_n, x) = \zeta_e(m(x_n, x)) < \epsilon$ for all $n \geq n_0$

(b) Let $\epsilon > 1$ be given. Since $\{x_n\}$ is a b-TVS cone multiplicative Cauchy sequence, it follows that there exists $n_0 \in \mathbb{N}$ such that $m(x_n, x_k) \ll e^\epsilon$ for all $n, k \geq n_0$. Thus, $m(x_n, x_k) \in \frac{e^\epsilon}{int(K)}$. Hence, from Lemma D.6(d), we see that $M(x_n, x_k) = \zeta_e(m(x_n, x_k)) < \epsilon$ for all $n, k \geq n_0$

(c) It follows from (a) and (b)

Theorem D.3 1

Let (X, m) be a b-TVS multiplicative cone metric space. Suppose mappings $f, g : X \mapsto X$ satisfy

(a) $f^r(X) \subseteq g^r(X)$ for any $r \in \mathbb{N}$ and $g^r(X)$ is a b-TVS cone multiplicative complete subspace of X for any $r \in \mathbb{N}$

(b) there exists $q \in [0, \frac{1}{a})$ and a modification on Proposition 1.11 [Ampadu, Clement (2016): Higher Order Banach Contraction Principle in Rectangular Multiplicative b-Metric Space, Unpublished] such that $m(f^r x, f^r y) \leq_K m(g^r x, g^r y)^{Zq^r}$ for all $x, y \in X$

Then f and g have a unique r-point of coincidence in X. Moreover, if f and g are r-weakly compatible, then f and g have a unique common r-fixed point.

Proof of Theorem D.3 1

Let $x_0 \in X$ be arbitrary. Choose a point $x_1 \in X$ such that $f^r x_0 = g^r x_1$. Continuing this process, having chosen $x_n \in X$, we obtain $x_{n+1} \in X$ such that $f^r(x_n) = g^r x_{n+1}$. Since $m(f^r x, f^r y) \leq_K m(g^r x, g^r y)^{Zq^r}$ we have that $m(g^r x, g^r y)^{Zq^r} \in K \cdot m(f^r x, f^r y)$ for all $x, y \in X$. It follows that $\zeta_e(m(f^r x, f^r y)) \leq \zeta_e(m(g^r x, g^r y)^{Zq^r}) \leq [\zeta_e(m(g^r x, g^r y))]^{Zq^r}$. Thus, $M(f^r x, f^r y) \leq M(g^r x, g^r y)^{Zq^r}$ for all $x, y \in X$. By induction we have, $M(g^r x_{n+1}, g^r x_n) \leq M(g^r x_1, g^r x_0)^{(Zq^r)^n}$. Thus, for all $p \geq 1$, we have that

$$M(g^r x_n, g^r x_{n+p}) \leq M(g^r x_n, g^r x_{n+1})^a \cdot M(g^r x_{n+1}, g^r x_{n+2})^{a^2} \cdots M(g^r x_{n+p-1}, g^r x_{n+p})^{a^p}$$
$$\leq M(g^r x_0, g^r x_1)^{a(Zq^r)^n + a^2(Zq^r)^{n+1} + \cdots + a^p(Zq^r)^{n+p+1}}$$
$$\leq M(g^r x_1, g^r x_0)^{\frac{a(Zq^r)^n}{1-Zq^r}}$$

Consequently, the sequence $\{g^r x_n\}$ is a multiplicative Cauchy sequence in the multiplicative b-metric space $(g^r X, M)$. Since $(g^r(X), M)$ is complete, there exists $q \in g^r(X)$ such that $g^r x_n \to q$ as $n \to \infty$. Thus, there exists $p \in X$ such that $g^r p = q$. Further, for each $\epsilon > 1$, there exists $n_0 \in \mathbb{N}$ such that for all $n \geq n_0$ we have $M(g^r x_n, f^r p) = M(f^r x_{n-1}, f^r p) \leq M(g^r x_{n-1}, g^r p)^{Zq^r} < \epsilon$. It follows that $g^r x_n \to f^r p$ as $n \to \infty$. Uniqueness of the limit implies that $f^r p = g^r p = q$. Now we show that f and g have a unique r-point of coincidence. Suppose there exists another point $p_1 \in X$ such that $f^r p_1 = g^r p_1$, then we have, $M(g^r p_1, g^r p) = M(f^r p_1, f^r p) \leq M(g^r p_1, g^r p)^{Zq^r}$. However, $1 - Zq^r \neq 0$. Hence, it follows that $M(g^r p_1, g^r p) = 1$, that is, $g^r p_1 = g^r p$. From Proposition D.12, f and g have a unique common r-fixed point.

If g is the identity in the previous theorem, we get the following

> **Corollary D.4 1**
>
> Let (X, m) be a complete b-TVS multiplicative cone metric space. Suppose mapping $f: X \mapsto X$ satisfies:
>
> there exists $q \in [0, \frac{1}{a})$ and Z given by Proposition 1.11 [Ampadu, Clement (2016): Higher Order Banach Contraction Principle in Rectangular Multiplicative b-Metric Space, Unpublished] such that $m(f^r x, f^r y) \leq_K m(x,y)^{Zq^r}$ for all $x, y \in X$
>
> Then f has a unique r-fixed point in X

4.4 Exercises

> **Exercise D.1 1**
>
> Verify Example D.8 and Example D.9

> **Exercise D.2 1**
>
> Let the setting be this chapter. Obtain the higher-order version of
>
> (a) Theorem 4 [Ion Marian Olaru, Adrian Branga, Anca Oprea, Common fixed point results in b-cone metric spaces over topological vector spaces, General Mathematics Vol. 20, No. 1 (2012), 57–67]
>
> (b) Theorem 5 [Ion Marian Olaru, Adrian Branga, Anca Oprea, Common fixed point results in b-cone metric spaces over topological vector spaces, General Mathematics Vol. 20, No. 1 (2012), 57–67]

> **Exercise D.3 1**
>
> Let the setting be this chapter. Deduce the following
>
> (a) the higher-order version of Corollary 2 [Ion Marian Olaru, Adrian Branga, Anca Oprea, Common fixed point results in b-cone metric spaces over topological vector spaces, General Mathematics Vol. 20, No. 1 (2012), 57–67] is a consequence of Exercise D.2(a)
>
> (b) the higher-order version of Corollary 3 [Ion Marian Olaru, Adrian Branga, Anca Oprea, Common fixed point results in b-cone metric spaces over topological vector spaces, General Mathematics Vol. 20, No. 1 (2012), 57–67] is a consequence of Exercise D.2(b)

> **Exercise D.4 1**
>
> Let ζ_e be the scalarization type function introduced in this chapter, and let (X, ρ) be a multiplicative cone b-metric space. Prove that the function $m_\rho : X \times X \mapsto [1, \infty)$ defined by $m_\rho = \zeta_e \circ \rho$ is a multiplicative b-metric, that is, obtain the multiplicative analogue of Corollary 2.2 [Wei-Shih Du and Erdal Karapınar, A note on cone b-metric and its related results: generalizations or equivalence?, Fixed Point Theory and Applications 2013, 2013:210]

> **Exercise D.5 1**
>
> Let the setting be this chapter. Obtain the higher-order version of Theorem 2.7 [Wei-Shih Du and Erdal Karapınar, A note on cone b-metric and its related results: generalizations or equivalence?, Fixed Point Theory and Applications 2013, 2013:210]

4.5 References

(1) M. Abbas, G. Jungck, Common fixed point results for non-commuting mappings without continuity in cone metric spaces, J. Math. Anal. Appl., 341, 2008

(2) Chen,G.Y., Huang,X.X., Yang, X.Q.: Vector Optimization, Springer-Verlag,Berlin, Heidelberg, Germany,2005

(3) Du Wei-Shih: On some nonlinear problems induced by an abstract maximal element principle, J. Math. Anal. Appl.,347 (2008), 391-399

(4) Du Wei-Shih, A note on cone metric fixed point theory and its equivalence, Nonlinear Analysis, 72 (5)(2010),2259-2261

(5) Ampadu, Clement (2016): Higher Order Banach Contraction Principle in Rectangular Multiplicative b-Metric Space, Unpublished

(6) Ion Marian Olaru, Adrian Branga, Anca Oprea, Common fixed point results in b-cone metric spaces over topological vector spaces, General Mathematics Vol. 20, No. 1 (2012), 57–67

(7) Wei-Shih Du and Erdal Karapınar, A note on cone b-metric and its related results: generalizations or equivalence?, Fixed Point Theory and Applications 2013, 2013:210

Chapter 5

r-Coincidence Point and r-Fixed Point Theorems in Multiplicative Cone b-Metric Space

5.1 Brief Summary

Abstract E.1 1

We obtain some coupled r-coincidence point results in multiplicative cone b-metric space as well as some common coupled r-fixed point results in multiplicative cone b-metric space.

5.2 Preliminaries

Notation E.1 1

E will denote a real Banach space and 1_E will denote the one element in E.

Definition E.2 1

A multiplicative cone P will be a subset of E such that

(a) P is nonempty, closed, and $P \neq \{1_E\}$

(b) if a, b are nonnegative real numbers and $x, y \in P$, then $x^a \cdot y^b \in P$

(c) $x \in P$ and $\frac{1}{x} \in P$ implies $x = 1_E$

Notation E.3 1

For any cone $P \subset E$, the partial ordering $\preceq$ with respect to P is defined as $x \preceq y$ iff $\frac{y}{x} \in P$; $\prec$ will mean $x \preceq y$ but $x \neq y$; $x \ll y$ will mean $\frac{y}{x} \in int(P)$, where $int(P)$ will denote the interior of P. The multiplicative cone P will be called multiplicative normal if there exists a number K such that $1_E \preceq x \preceq y$ implies $\|x\| \leq \|y\|^K$ for all $x, y \in E$. The least positive number K for which $\|x\| \leq \|y\|^K$ holds will be called the multiplicative normal constant of P

Remark E.4 1

We assume that $int(P) \neq \emptyset$

CHAPTER 5. R-COINCIDENCE POINT AND R-FIXED POINT THEOREMS IN MULTIPLICATIVE CONE B-METRIC SPACE

Definition E.5 1

Let X be a nonempty set and E be a real Banach space equipped with partial order $\preceq$ with respect to the multiplicative cone P. A function $m : X \times X \mapsto E$ will be called a multiplicative cone b-metric on X with constant $s \geq 1$ if the following conditions are satisfied

(a) $1_E \preceq m(x,y)$ for all $x,y \in X$ and $m(x,y) = 1_E$ iff $x = y$

(b) $m(x,y) = m(y,x)$ for all $x,y \in X$

(c) $m(x,y) \preceq [m(x,z) \cdot m(z,y)]^s$ for all $x,y,z \in X$

When m satisfies the above conditions we say (X, m) is a multiplicative cone b-metric space

Example E.6 1

Let $X = \{-1, 0, 1\}$, $E = \mathbb{R}^2$, $P = \{(x,y) : x \geq 1, y \geq 1\}$. Define $m : X \times X \mapsto P$ by $m(x,y) = m(y,x)$ for all $x,y \in X$; $m(x,x) = 1_E$, $x \in X$; $m(-1,0) = (a^3, a^3)$ for some $a > 1$; $m(-1,1) = m(0,1) = (a,a)$ for some $a > 1$. Then (X,m) is a multiplicative cone b-metric space. Note that the multiplicative triangle inequality does not hold since $m(-1,1) \cdot m(1,0) = (a,a) \cdot (a,a) = (a^2, a^2) \prec (a^3, a^3) = m(-1,0)$. Note that the co-efficient is given by $s = \frac{3}{2}$

Example E.7 1

Let $X = \mathbb{N} \cup \{\infty\}$, $E = \mathbb{R}^2$ and $P = \{(x,y) \in E : x \geq 1, y \geq 1\}$. Define $m : X \times X \mapsto E$ by, $m(x,y) = (1,1)$ if $x = y$; $m(x,y) = (a^{|\frac{1}{x} - \frac{1}{y}|}, a^{|\frac{1}{x} - \frac{1}{y}|})$ for some $a > 1$ and if x and y are even or $xy = \infty$; $m(x,y) = (a^5, a^5)$ for some $a > 1$ and if x and y are odd and $x \neq y$; $m(x,y) = (a^2, a^2)$ for some $a > 1$ and otherwise. Then (X,m) is a multiplicative cone b-metric space with coefficient $s = 3$. Note that (X,m) is not a multiplicative cone metric space since the multiplicative triangle inequality does not hold. Indeed, for any $a > 1$, we have, $(a^5, a^5) = m(1,3) \succ m(1,2) \cdot m(2,3) = (a^2, a^2) \cdot (a^2, a^2) = (a^4, a^4)$

Definition E.8 1

Let (X,m) be a multiplicative cone b-metric space, and let $\{x_n\}$ be a sequence in X and $x \in X$

(a) For all $c \in E$ with $1 \ll c$, if there exists a positive integer N such that $m(x_n, x) \ll c$ for all $n > N$, then we will say x_n is multiplicative convergent and x is the limit of $\{x_n\}$

(b) For all $c \in E$ with $1 \ll c$, if there exists a positive integer N such that $m(x_n, x_k) \ll c$ for all $n, k > N$, then we will say that $\{x_n\}$ is a multiplicative Cauchy sequence in X

(c) The multiplicative cone b-metric space (X, m) will be called multiplicative complete if every multiplicative Cauchy sequence in X is multiplicative convergent.

Taking inspiration from [Jungck, G, Radenovic, S, Radojevic, S, Rakocevic, V: Common fixed point theorems for weakly compatible pairs on cone metric spaces. Fixed Point Theory Appl. 2009, Article ID 643840 (2009)], we have the following

Lemma E.9 1

(a) If E is a real Banach space with multiplicative cone P and $a \preceq a^\lambda$, where $a \in P$ and $0 \leq \lambda < 1$, then $a = 1$

(b) If $c \in int(P)$, $1_E \preceq a_n$ and $a_n \to 1_E$, then there exists a positive integer N such that $a_n \ll c$ for all $n \geq N$

(c) If $a \preceq b$ and $b \ll c$, then $a \ll c$

(d) If $1 \preceq u \ll c$ for each $1 \ll c$, then $u = 1_E$

Taking inspiration from [Bhaskar, TG, Lakshmikantham, V: Fixed point theorems in partially ordered metric spaces and applications. Nonlinear Anal., Theory Methods Appl. 65(7), 1379-1393 (2006)] we introduce the following

Definition E.10 1

An element $(x,y) \in X^2$ will be called a coupled r-fixed point of the mapping $F : X^2 \mapsto X$ if $F^r(x,y) = x$ and $F^r(y,x) = y$ for any $r \in \mathbb{N}$

Taking inspiration from [Lakshmikantham, V, Ciric, L: Coupled fixed point theorems for nonlinear contractions in partially ordered metric spaces. Nonlinear Anal., Theory Methods Appl. 70(12), 4341-4349 (2009)] we introduce the following

Definition E.11 1

An element $(x,y) \in X^2$ will be called

(a) a coupled r-coincidence point of the mappings $F : X^2 \mapsto X$ and $g : X \mapsto X$ if $g^r x = F^r(x,y)$ and $g^r y = F^r(y,x)$ for any $r \in \mathbb{N}$. Moreover, in this instance, $(g^r x, g^r y)$ for any $r \in \mathbb{N}$ will be called a coupled r-point of coincidence

(b) a common coupled r-fixed point of the mappings $F : X^2 \mapsto X$ and $g : X \mapsto X$ if $x = g^r x = F^r(x,y)$ and $y = g^r y = F^r(y,x)$ for any $r \in \mathbb{N}$

Taking inspiration from [Abbas, M, Khan, MA, Radenovic, S: Common coupled fixed point theorems in cone metric spaces for w-compatible mappings. Appl. Math. Comput. 217, 195-202 (2010)] we introduce the following

Definition E.12 1

The mappings $F : X^2 \mapsto X$ and $g : X \mapsto X$ will be called r-w-compatible if $g^r(F^r(x,y)) = F^r(g^r x, g^r y)$ whenever $g^r x = F^r(x,y)$ and $g^r y = F^r(y,x)$ for any $r \in \mathbb{N}$

Definition E.13 1

Let (X,d) be a cone b-metric space with coefficient $s \geq 1$. The mappings $F : X^2 \mapsto X$ and $g : X \mapsto X$ will be said to form an (a,b)-contraction if $d(F(x,y), F(u,v)) \leq ad(gx, gu) + bd(gy, gv)$ holds for all $x, y, u, v \in X$, where $a + b < \frac{1}{s}$ and $a, b \geq 0$

Definition E.14 1

Let (X,d) be a cone b-metric space with coefficient $s \geq 1$. The mappings $F : X^2 \mapsto X$ and $g : X \mapsto X$ will be said to form an (a,b)- type-contraction if $d(F(x,y), F(u,v)) \leq k[d(gx, gu) + d(gy, gv)]$ holds for all $x, y, u, v \in X$, where $k < \frac{1}{2s}$ and k is nonnegative

CHAPTER 5. R-COINCIDENCE POINT AND R-FIXED POINT THEOREMS IN MULTIPLICATIVE CONE B-METRIC SPACE

Definition E.15 1

Let (X, d) be a cone b-metric space with coefficient $s \geq 1$. The mappings $F : X^2 \mapsto X$ and $g : X \mapsto X$ will be said to form a higher-order (a, b)-type-contraction if $d(F^r(x, y), F^r(u, v)) \leq \sum_{q=0}^{r-1} c_q [d(g^{q+1}v, g^{q+1}u) + d(g^{q+1}y, g^{q+1}v)]$ holds for all $x, y, u, v \in X$, and $r \in \mathbb{N}$, where $0 \leq c_q < \frac{1}{2s}$ for all $0 \leq q \leq r-1$

Proposition E.16 1

Let (X, d) be a cone b-metric space with coefficient $s \geq 1$, and let the mappings $F : X^2 \mapsto X$ and $g : X \mapsto X$ form a higher-order (a, b)-type-contraction. For every pair $x \neq y \in X$ and $y \neq v \in X$, define

$$M := M(x, y) = \max_{0 \leq k \leq r-1} q^{-k} \frac{d(F^k(x, y), F^k(y, v))}{d(gx, gu) + d(gy, gu)}$$

then

$$M = \max_{n \in \{0\} \cup \mathbb{N}} q^{-n} \frac{d(F^n(x, y), F^n(y, v))}{d(gx, gu) + d(gy, gu)}$$

, where $q \in [0, \frac{1}{2s})$

Now we have the following alternate characterization of higher-order (a, b)-type-contraction

Definition E.17 1

Let (X, d) be a cone b-metric space with coefficient $s \geq 1$. The mappings $F : X^2 \mapsto X$ and $g : X \mapsto X$ will be said to form a higher-order (a, b)-type-contraction if $d(F^r(x, y), F^r(u, v)) \leq Mq^r[d(gx, gu) + d(gy, gu)]$ for all $x, y, u, v \in X$, where M and q are given by the previous proposition.

5.3 Main Results

Theorem E.1 1

Let (X, m) be a multiplicative cone b-metric space with coefficient $s \geq 1$ relative to a solid multiplicative cone P. Let $F : X^2 \mapsto X$ and $g : X \mapsto X$ satisfy $m(F^r(x, y), F^r(u, v)) \leq [m(g^r x, g^r u) \cdot m(g^r y, g^r v)]^{M^\star q^r}$ for all $x, y, u, v \in X$, where q is given by the previous proposition but $M^\star$ is a certain modification on it. If $F^r(X^2) \subseteq g^r(X)$ for any $r \in \mathbb{N}$ and $g^r(X)$ is a complete subspace of X for any $r \in \mathbb{N}$, then F and g have a coupled r-coincidence point $(x^\star, y^\star) \in X^2$

> **Proof of Theorem E.1 1**
>
> Choose $x_0, y_0 \in X$, and set $g^r x_1 = F^r(x_0, y_0)$, $g^r y_1 = F^r(y_0, x_0)$, this can be done since $F^r(X^2) \subseteq g^r(X)$. Continuing this process, we obtain two sequences $\{x_n\}$ and $\{y_n\}$ such that $g^r x_{n+1} = F^r(x_n, y_n)$ and $g^r y_{n+1} = F^r(y_n, x_n)$. Now
>
> $$m(g^r x_n, g^r x_{n+1}) = m(F^r(x_{n-1}, y_{n-1}), F^r(x_n, y_n)) \leq [m(g^r x_{n-1}, g^r x_n) \cdot m(g^r y_{n-1}, g^r y_n)]^{M^\star q^r}$$
>
> On the other hand
>
> $$m(g^r y_n, g^r y_{n+1}) \leq [m(g^r y_{n-1}, g^r y_n) \cdot m(g^r x_{n-1}, g^r x_n)]^{M^\star q^r}$$
>
> Now set $b_n = m(g^r x_n, g^r x_{n+1}) \cdot m(g^r y_n, g^r y_{n+1})$, then it follows from the two inequalities immediately above that
>
> $$b_n \leq b_{n-1}^{2M^\star q^r}$$
>
> and by induction we have $b_n \leq b_0^{(2M^\star q^r)^n}$, where $2M^\star q^r < \frac{1}{s}$. Now let $k > n \geq 1$, then,
>
> $$m(g^r x_n, g^r x_k) \leq m(g^r x_n, g^r x_{n+1})^s \cdot m(g^r x_{n+1}, g^r x_{n+2})^{s^2} \cdot \ldots \cdot m(g^r x_{k-1}, g^r x_k)^{s^{k-n}}$$
>
> and
>
> $$m(g^r y_n, g^r y_k) \leq m(g^r y_n, g^r y_{n+1})^s \cdot m(g^r y_{n+1}, g^r y_{n+2})^{s^2} \cdot \ldots \cdot m(g^r y_{k-1}, g^r y_k)^{s^{k-n}}$$
>
> From the above two inequalities and the fact that, $b_n \leq b_0^{(2M^\star q^r)^n}$, we have, with $h = 2M^\star q^r$, that,
>
> $$m(g^r x_n, g^r x_k) \cdot m(g^r y_n, g^r y_k) \leq b_n^s \cdot b_{n+1}^{s^2} \cdot \ldots \cdot b_{k-1}^{s^{k-n}}$$
> $$\leq b_0^{sh^n + s^2 h^{n+1} + \cdots + s^{k-n} h^{k-1}}$$
> $$\leq b_0^{\frac{sh^n}{1-sh}} \to 1 \text{ as } n \to \infty$$
>
> For any $c \in E$ with $1_E \ll c$, according to Lemma E.9, there exists $N_0 \in \mathbb{N}$ such that $b_0^{\frac{sh^n}{1-sh}} \ll c$. It follows from the above "chain of inequalities" and Lemma E.9, that for any $k > n > N_0$, we have, $m(g^r x_n, g^r x_k) \cdot m(g^r y_n, g^r y_k) \ll c$, and thus, $m(g^r x_n, g^r x_k) \ll c$ and $m(g^r y_n, g^r y_k) \ll c$. It follows that $\{g^r x_n\}$ and $\{g^r y_n\}$ are multiplicative Cauchy sequences in $g^r(X)$. Since $g^r(X)$ is complete, there exists $x^\star, y^\star \in X$ such that $g^r x_n \to g^r x^\star$ and $g^r y_n \to g^r y^\star$ as $n \to \infty$. Now notice that,
>
> $$m(F^r(x^\star, y^\star), g^r x^\star) \leq [m(F^r(x^\star, y^\star), g^r x_{n+1}) \cdot m(g^r x_{n+1}, g^r x^\star)]^s$$
> $$= [m(F^r(x^\star, y^\star), F^r(x_n, y_n)) \cdot m(g^r x_{n+1}, g^r x^\star)]^s$$
> $$\leq [m(g^r x^\star, g^r x_n) \cdot m(g^r y^\star, g^r y_n)]^{sM^\star q^r} \cdot m(g^r x_{n+1}, g^r x^\star)^s$$
>
> Similarly, we have,
>
> $$m(F^r(y^\star, x^\star), g^r y^\star) \leq [m(g^r y^\star, g^r y_n) \cdot m(g^r x^\star, g^r x_n)]^{sM^\star q^r} \cdot m(g^r y_{n+1}, g^r y^\star)^s$$
>
> Now put $\delta = m(F^r(x^\star, y^\star), g^r x^\star) \cdot m(F^r(y^\star, x^\star), g^r y^\star)$, then it follows from the two inequalities immediately above that,
>
> $$\delta \leq [m(g^r x^\star, g^r x_n) \cdot m(g^r y^\star, g^r y_n)]^{2sM^\star q^r} \cdot [m(g^r x_{n+1}, g^r x^\star) \cdot m(g^r y_{n+1}, g^r y^\star)]^s$$

Proof of Theorem E.1 continued 1

Since, $g^r x_n \to g^r x^\star$ and $g^r y_n \to g^r y^\star$ as $n \to \infty$, then by Definition E.8, and for $1_E \ll c$, there exists $N_0 \in \mathbb{N}$ such that for all $n > N_0$, we have, $m(g^r x^\star, g^r x_n) \ll c^{\frac{1}{8sM^\star q^r}}$, $m(g^r y^\star, g^r y_n) \ll c^{\frac{1}{8sM^\star q^r}}$, $m(g^r x_{n+1}, g^r x^\star) \ll c^{\frac{1}{4s}}$, $m(g^r y_{n+1}, g^r y^\star) \ll c^{\frac{1}{4s}}$. Thus, it follows that $\delta \ll [c^{\frac{1}{8sM^\star q^r}} \cdot c^{\frac{1}{8sM^\star q^r}}]^{2sM^\star q^r} \cdot [c^{\frac{1}{4s}} \cdot c^{\frac{1}{4s}}]^s = c^{\frac{1}{4}} \cdot c^{\frac{1}{4}} \cdot c^{\frac{1}{4}} \cdot c^{\frac{1}{4}} = c$. Thus, according to Lemma E.9, $\delta = 1_E$, thus it follows that, $m(F^r(x^\star, y^\star), g^r x^\star) = 1_E$ and $m(F^r(y^\star, x^\star), g^r y^\star) = 1_E$, hence, $F^r(x^\star, y^\star) = g^r x^\star$ and $F^r(y^\star, x^\star) = g^r y^\star$, that is, $(x^\star, y^\star)$ is a coupled r-coincidence point of F and g

Theorem E.2 1

In addition to the hypotheses of the previous theorem, if F and g are r-w-compatible, then F and g have a unique common coupled r-fixed point. Moreover, a common coupled r-fixed point of F and g is of the form (u, u) for some $u \in X$

Proof of Theorem E.2 1

From the previous theorem, F and g have a coupled r-coincidence point $(x^\star, y^\star)$. It follows that $(g^r x^\star, g^r y^\star)$ is a coupled r-point of coincidence of F and g such that $g^r x^\star = F^r(x^\star, y^\star)$ and $g^r y^\star = F^r(y^\star, x^\star)$. Now we show the coupled r-point of coincidence is unique. Suppose that F and g have another coupled r-point of coincidence $(g^r x', g^r y')$ such that $g^r x' = F^r(x', y')$ and $g^r y' = F^r(y', x')$, where $(x', y') \in X^2$. Then we have,

$$m(g^r x^\star, g^r x') = m(F^r(x^\star, y^\star), F^r(x', y')) \leq [m(g^r x^\star, g^r x') \cdot m(g^r y^\star, g^r y')]^{M^\star q^r}$$

Similarly, we obtain,

$$m(g^r y^\star, g^r y') = m(F^r(y^\star, x^\star), F^r(y', x')) \leq [m(g^r y^\star, g^r y') \cdot m(g^r x^\star, g^r x')]^{M^\star q^r}$$

Put $\gamma := m(g^r y^\star, g^r y') \cdot m(g^r x^\star, g^r x')$, then it follows from the two inequalities immediately above that $\gamma \leq \gamma^{2M^\star q^r}$, but $1 - 2M^\star q^r \neq 0$, thus, $\gamma = 1$ by Lemma E.9, and since $m(x, y) \geq 1$, it follows that $m(g^r y^\star, g^r y') = 1_E$ and $m(g^r x^\star, g^r x') = 1_E$, that is, $g^r y^\star = g^r y'$ and $g^r x^\star = g^r x'$. It follows that the r-coupled point of coincidence of F and g is unique. Similarly, we can show that $g^r x^\star = g^r x'$ and $g^r y^\star = g^r y'$. Since we can conclude $g^r x^\star = g^r y^\star$, it follows that the unique coupled r-point of coincidence of F and g is $(g^r x^\star, g^r x^\star)$. Now let $u = g^r x^\star = F^r(x^\star, y^\star)$. Since F and g are r-w-compatible, then we have $g^r u = g^r(g^r x^\star) = g^r F^r(x^\star, y^\star) = F^r(g^r x^\star, g^r y^\star) = F^r(u, u)$. Thus, $(g^r u, g^r u)$ is a coupled r-point of coincidence, and also (u, u) is a coupled r-point of coincidence. The uniqueness of the coupled r-point of coincidence implies that $g^r u = u$. Therefore $u = g^r u = F^r(u, u)$. Hence, (u, u) is the unique common coupled r-fixed point of F and g

As a consequence of the previous two theorems the following are immediate

Corollary E.3 1

Let (X, m) be a multiplicative cone b-metric space with coefficient $s \geq 1$ relative to a solid multiplicative cone P. Let $F : X^2 \mapsto X$ and $g : X \mapsto X$ satisfy $m(F^r(x, y), F^r(u, v)) \leq [m(g^r x, F^r(x, y)) \cdot m(g^r u, F^r(u, v))]^{M^{\star\star} q^r}$ for all $x, y, u, v \in X$, where q is given by the previous proposition but $M^{\star\star}$ is a certain modification on it. If $F^r(X^2) \subseteq g^r(X)$ for any $r \in \mathbb{N}$ and $g^r(X)$ is a complete subspace of X for any $r \in \mathbb{N}$, then F and g have a coupled r-coincidence point $(x^\star, y^\star) \in X^2$

CHAPTER 5. R-COINCIDENCE POINT AND R-FIXED POINT THEOREMS IN MULTIPLICATIVE CONE B-METRIC SPACE

Corollary E.4 1

Let (X, m) be a multiplicative cone b-metric space with coefficient $s \geq 1$ relative to a solid multiplicative cone P. Let $F : X^2 \mapsto X$ and $g : X \mapsto X$ satisfy $m(F^r(x,y), F^r(u,v)) \leq [m(g^r x, F^r(u,v)) \cdot m(g^r u, F^r(x,y))]^{M^{\star\star\star} q^r}$ for all $x, y, u, v \in X$, where q is given by the previous proposition but $M^{\star\star\star}$ is a certain modification on it. If $F^r(X^2) \subseteq g^r(X)$ for any $r \in \mathbb{N}$ and $g^r(X)$ is a complete subspace of X for any $r \in \mathbb{N}$, then F and g have a coupled r-coincidence point $(x^\star, y^\star) \in X^2$

Corollary E.5 1

Let (X, m) be a multiplicative cone b-metric space with coefficient $s \geq 1$ relative to a solid multiplicative cone P. Let $F : X^2 \mapsto X$ satisfy $m(F^r(x,y), F^r(u,v)) \leq [m(x,u) \cdot m(y,v)]^{M^{\bullet} q^r}$ for all $x, y, u, v \in X$, where q is given by the previous proposition but $M^{\bullet}$ is a certain modification on it. Then F has a coupled r-fixed point $(x^\star, y^\star) \in X^2$. Moreover, the coupled r-fixed point is unique and it is of the form $(x^\star, x^\star)$ for some $x^\star \in X$

Corollary E.6 1

Let (X, m) be a multiplicative cone b-metric space with coefficient $s \geq 1$ relative to a solid multiplicative cone P. Let $F : X^2 \mapsto X$ satisfy $m(F^r(x,y), F^r(u,v)) \leq [m(x, F(x,y)) \cdot m(u, F(u,v))]^{M^{\bullet\bullet} q^r}$ for all $x, y, u, v \in X$, where q is given by the previous proposition but $M^{\bullet\bullet}$ is a certain modification on it. Then F has a coupled r-fixed point $(x^\star, y^\star) \in X^2$. Moreover, the coupled r-fixed point is unique and it is of the form $(x^\star, x^\star)$ for some $x^\star \in X$

Corollary E.7 1

Let (X, m) be a multiplicative cone b-metric space with coefficient $s \geq 1$ relative to a solid multiplicative cone P. Let $F : X^2 \mapsto X$ satisfy $m(F^r(x,y), F^r(u,v)) \leq [m(x, F(u,v)) \cdot m(u, F(x,y))]^{M^{\bullet\bullet\bullet} q^r}$ for all $x, y, u, v \in X$, where q is given by the previous proposition but $M^{\bullet\bullet\bullet}$ is a certain modification on it. Then F has a coupled r-fixed point $(x^\star, y^\star) \in X^2$. Moreover, the coupled r-fixed point is unique and it is of the form $(x^\star, x^\star)$ for some $x^\star \in X$

5.4 Exercises

Exercise E.1 1

(a) Verify Example E.6, that is, show (X, m) is a multiplicative cone b-metric space with coefficient $s = \frac{3}{2}$

(b) Verify Example E.7, that is, show (X, m) is a multiplicative cone b-metric space with coefficient $s = 3$

Exercise E.2 1

Let the setting be this chapter. Obtain the higher-order version of the following

(a) Theorem 2.1 [Fadail and Ahmad Fixed Point Theory and Applications 2013, 2013:177]

(b) Theorem 3.1 [Fadail and Ahmad Fixed Point Theory and Applications 2013, 2013:177]

Exercise E.3 1

(a) Show that Theorem E.1 is a special case of Exercise E.2(a)

(b) Show that Theorem E.2 is a special case of Exercise E.2(b)

> **Exercise E.4 1**
>
> Verify the following
>
> (a) Corollary E.3 is the higher-order version of Corollary 2.4 [Fadail and Ahmad Fixed Point Theory and Applications 2013, 2013:177]
>
> (b) Corollary E.4 is the higher-order version of Corollary 2.5 [Fadail and Ahmad Fixed Point Theory and Applications 2013, 2013:177]

> **Exercise E.5 1**
>
> Obtain the higher-order version of Theorem 3.3 [Fadail and Ahmad Fixed Point Theory and Applications 2013, 2013:177]

> **Exercise E.6 1**
>
> (a) Show that Corollary E.5 is a special case of Exercise E.5
>
> (b) Show that Corollary E.6 is the higher-order version of Corollary 3.5 [Fadail and Ahmad Fixed Point Theory and Applications 2013, 2013:177]
>
> (c) Show that Corollary E.7 is the higher-order version of Corollary 3.6 [Fadail and Ahmad Fixed Point Theory and Applications 2013, 2013:177]

5.5 References

(1) Jungck, G, Radenovic, S, Radojevic, S, Rakocevic, V: Common fixed point theorems for weakly compatible pairs on cone metric spaces. Fixed Point Theory Appl. 2009, Article ID 643840 (2009)

(2) Bhaskar, TG, Lakshmikantham, V: Fixed point theorems in partially ordered metric spaces and applications. Nonlinear Anal., Theory Methods Appl. 65(7), 1379-1393 (2006)

(3) Lakshmikantham, V, Ciric, L: Coupled fixed point theorems for nonlinear contractions in partially ordered metric spaces. Nonlinear Anal., Theory Methods Appl. 70(12) 4341-4349 (2009)

(4) Abbas, M, Khan, MA, Radenovic, S: Common coupled fixed point theorems in cone metric spaces for w-compatible mappings. Appl. Math. Comput. 217, 195-202 (2010)

(5) Fadail and Ahmad Fixed Point Theory and Applications 2013, 2013:177

Chapter 6

r-g-Monotone and r-w-Compatible Mappings in Ordered Multiplicative Cone b-Metric Space

6.1 Brief Summary

> **Abstract F.1 1**
>
> In this chapter we obtain some coupled r-coincidence point and common coupled r-fixed point results in ordered multiplicative cone b-metric space.

6.2 Preliminaries

> **Notation F.1 1**
>
> E will denote a real Banach space and 1_E will denote the one element in E

> **Definition F.2 1**
>
> A multiplicative cone P will be a subset of E such that
>
> (a) P is nonempty, closed, and $P \neq \{1_E\}$
>
> (b) if a, b are nonnegative real numbers and $x, y \in P$, then $x^a \cdot y^b \in P$
>
> (c) $x \in P$ and $\frac{1}{x} \in P$ implies $x = 1_E$

> **Notation F.3 1**
>
> For any cone $P \subset E$, the partial ordering $\preceq$ with respect to P is defined as $x \preceq y$ iff $\frac{y}{x} \in P$; $\prec$ will mean $x \preceq y$ but $x \neq y$; $x \ll y$ will mean $\frac{y}{x} \in int(P)$, where $int(P)$ will denote the interior of P. The multiplicative cone P will be called multiplicative normal if there exists a number K such that $1_E \preceq x \preceq y$ implies $\|x\| \leq \|y\|^K$ for all $x, y \in E$. The least positive number K for which $\|x\| \leq \|y\|^K$ holds will be called the multiplicative normal constant of P

CHAPTER 6. R-G-MONOTONE AND R-W-COMPATIBLE MAPPINGS IN ORDERED MULTIPLICATIVE CONE B-METRIC SPACE

Remark F.4 1

We assume that $int(P) \neq \emptyset$

Definition F.5 1

Let X be a nonempty set and E be a real Banach space equipped with partial order $\preceq$ with respect to the multiplicative cone P. A function $m : X \times X \mapsto E$ will be called a multiplicative cone b-metric on X with constant $s \geq 1$ if the following conditions are satisfied

(a) $1_E \preceq m(x,y)$ for all $x, y \in X$ and $m(x,y) = 1_E$ iff $x = y$

(b) $m(x,y) = m(y,x)$ for all $x, y \in X$

(c) $m(x,y) \preceq [m(x,z) \cdot m(z,y)]^s$ for all $x, y, z \in X$

When m satisfies the above conditions we say (X, m) is a multiplicative cone b-metric space

Definition F.6 1

Let (X, m) be a multiplicative cone b-metric space, and let $\{x_n\}$ be a sequence in X and $x \in X$

(a) For all $c \in E$ with $1 \ll c$, if there exists a positive integer N such that $m(x_n, x) \ll c$ for all $n > N$, then we will say x_n is multiplicative convergent and x is the limit of $\{x_n\}$

(b) For all $c \in E$ with $1 \ll c$, if there exists a positive integer N such that $m(x_n, x_k) \ll c$ for all $n, k > N$, then we will say that $\{x_n\}$ is a multiplicative Cauchy sequence in X

(c) The multiplicative cone b-metric space (X, m) will be called multiplicative complete if every multiplicative Cauchy sequence in X is multiplicative convergent.

Taking inspiration from [Jungck, G, Radenovic, S, Radojevic, S, Rakocevic, V: Common fixed point theorems for weakly compatible pairs on cone metric spaces. Fixed Point Theory Appl. 2009, Article ID 643840 (2009)], we have the following

Lemma F.7 1

(a) If E is a real Banach space with multiplicative cone P and $a \preceq a^\lambda$, where $a \in P$ and $0 \leq \lambda < 1$, then $a = 1$

(b) If $c \in int(P)$, $1_E \preceq a_n$ and $a_n \to 1_E$, then there exists a positive integer N such that $a_n \ll c$ for all $n \geq N$

(c) If $a \preceq b$ and $b \ll c$, then $a \ll c$

(d) If $1 \preceq u \ll c$ for each $1 \ll c$, then $u = 1_E$

Taking inspiration from [Bhaskar, T.G., Lakshmikantham, V.: Fixed point theorems in partially ordered metric spaces and applications. Nonlinear Anal. 65(7), 1379–1393 (2006)] we introduce the following

Definition F.8 1

Let $(X, \sqsubseteq)$ be a partially ordered set and let $F : X^2 \mapsto X$ and $g : X \mapsto X$ be two mappings. The mapping F will be said to have the mixed r-g-monotone property if F is monotone r-g-non-decreasing in its first argument and monotone r-g-non-decreasing in its second argument, that is, for any $x, y \in X$, $x_1, x_2 \in X$, $g^r x_1 \sqsubseteq g^r x_2$ implies that $F^r(x_1, y) \sqsubseteq F^r(x_2, y)$ for any $r \in \mathbb{N}$; $y_1, y_2 \in X$, $g^r y_1 \sqsubseteq g^r y_2$ implies that $F^r(x, y_1) \sqsupseteq F^r(x, y_2)$ for any $r \in \mathbb{N}$

Taking inspiration from [Bhaskar, T.G., Lakshmikantham, V.: Fixed point theorems in partially ordered metric spaces and applications. Nonlinear Anal. 65(7), 1379–1393 (2006)] we introduce the following

Definition F.9 1

An element $(x,y) \in X^2$ will be called a coupled r-fixed point of the mapping $F : X^2 \mapsto X$ if $F^r(x,y) = x$ and $F^r(y,x) = y$ for any $r \in \mathbb{N}$

Taking inspiration from [Lakshmikantham, V., Ciric, L.: Coupled fixed point theorems for nonlinear contractions in partially ordered metric spaces. Nonlinear Anal. 70(12), 4341–4349 (2009)] we introduce the following

Definition F.10 1

An element $(x,y) \in X^2$ will be called

(a) a coupled r-coincidence point of the mappings $F : X^2 \mapsto X$ and $g : X \mapsto X$ if $g^r x = F^r(x,y)$ and $g^r y = F^r(y,x)$ for any $r \in \mathbb{N}$. Moreover, we say $(g^r x, g^r y)$ is a coupled r-point of coincidence for any $r \in \mathbb{N}$

(b) a common coupled r-fixed point of the mappings $F : X^2 \mapsto X$ and $g : X \mapsto X$ if $x = g^r x = F^r(x,y)$ and $y = g^r y = F^r(y,x)$ for any $r \in \mathbb{N}$

Taking inspiration from [Abbas, M., Khan, M.A., Radenovic, S.: Common coupled fixed point theorems in cone metric spaces for w-compatible mappings. Appl. Math. Comput. 217, 195–202 (2010)] we introduce the following

Definition F.11 1

The mappings $F : X^2 \mapsto X$ and $g : X \mapsto X$ will be called

(a) r-w-compatible for any $r \in \mathbb{N}$, if $g^r(F^r(x,y)) = F^r(g^r x, g^r y)$ whenever $g^r x = F^r(x,y)$ and $g^r y = F^r(y,x)$

(b) r-$w^\star$-compatible for any $r \in \mathbb{N}$, if $g^r(F^r(x,x)) = F^r(g^r x, g^r x)$

6.3 Main Result

Theorem F.1 1

Let $(X, \sqsubseteq)$ be a partially ordered set and (X, m) be a multiplicative cone b-metric space with coefficient $s \geq 1$ relative to a solid multiplicative cone P. Let $F : X^2 \mapsto X$ and $g : X \mapsto X$ be two mappings such that F has the mixed r-g-monotone property. Suppose for all $(x,y),(u,v) \in X^2$ with $g^r u \sqsubseteq g^r x$ and $g^r v \sqsupseteq g^r y$ or $g^r x \sqsubseteq g^r u$ and $g^r y \sqsupseteq g^r v$ it holds that $m(F^r(x,y), F^r(u,v)) \leq [m(g^r x, g^r u) \cdot m(g^r y, g^r v)]^{M^\star q^r}$, where q is given by Proposition 1.16 [Ampadu, Clement (2016): r-Coincidence Point and r-Fixed Point Theorems in Multiplicative Cone b-Metric Space, Unpublished] but $M^\star$ is a certain modification on the same Proposition. Assume that F and g satisfy the following conditions

(a) $F^r(X^2) \subseteq g^r(X)$ for any $r \in \mathbb{N}$

(b) $g^r(X)$ is a complete subspace of X for any $r \in \mathbb{N}$

> **Theorem F.1 Continued 1**
>
> Also suppose that X has the following properties
>
> (c) if a non-decreasing sequence $\{x_n\}$ in X is such that $x_n \to x$, then $x_n \sqsubseteq x$ for all $x \in \mathbb{N}$
>
> (d) if a non-increasing sequence $\{y_n\}$ in X is such that $y_n \to y$, then $y_n \sqsupseteq y$ for all $n \in \mathbb{N}$
>
> If there exist $x_0, y_0 \in X$ such that $g^r x_0 \sqsubseteq F^r(x_0, y_0)$ and $F^r(y_0, x_0) \sqsubseteq g^r y_0$, then F and g have a coupled r-coincidence point $(x^\star, y^\star) \in X^2$

Proof of Theorem F.1 1

Let $x_0, y_0 \in X$ be such that $g^r x_0 \sqsubseteq F^r(x_0, y_0)$ and $F^r(y_0, x_0) \sqsubseteq g^r y_0$. Since $F^r(X^2) \subseteq g^r(X)$ we can choose $x_1, y_1 \in X$ such that $g^r x_1 = F^r(x_0, y_0)$ and $g^r y_1 = F^r(y_0, x_0)$. Again since $F^r(X^2) \subseteq g^r(X)$, we can choose $x_2, y_2 \in X$ such that $g^r x_2 = F^r(x_1, y_1)$ and $g^r y_2 = F^r(y_1, x_1)$. Since F has the mixed r-g-monotone property, we have, $g^r x_0 \sqsubseteq g^r x_1 \sqsubseteq g^r x_2$ and $g^r y_2 \sqsubseteq g^r y_1 \sqsubseteq g^r y_0$. Continuing this process, we can construct two sequences $\{x_n\}$ and $\{y_n\}$ in X such that $g^r x_n = F^r(x_{n-1}, y_{n-1}) \sqsubseteq g^r x_{n+1} = F^r(x_n, y_n)$ and $g^r y_{n+1} = F^r(y_n, x_n) \sqsubseteq g^r y_n = F^r(y_{n-1}, x_{n-1})$. Now notice that

$$m(g^r x_n, g^r x_{n+1}) = m(F^r(x_{n-1}, y_{n-1}), F^r(x_n, y_n)) \preceq [m(g^r x_{n-1}, g^r x_n) \cdot m(g^r y_{n-1}, g^r y_n)]^{M^\star q^r}$$

and similarly, we have,

$$m(g^r y_n, g^r y_{n+1}) = m(F^r(y_{n-1}, x_{n-1}), F^r(y_n, x_n)) \preceq [m(g^r y_{n-1}, g^r y_n) \cdot m(g^r x_{n-1}, g^r x_n)]^{M^\star q^r}$$

Now put $\delta_n = m(g^r x_n, g^r x_{n+1}) \cdot m(g^r y_n, g^r y_{n+1})$, then it follows that $\delta_n \preceq \delta_{n-1}^{2M^\star q^r}$. Now put $h = 2M^\star q^r < \frac{1}{s}$, then it follows that $\delta_n \preceq \delta_{n-1}^h$, and by induction $\delta_n \preceq \delta_0^{h^n}$. Now let $k > n \geq 1$, then,

$$m(g^r x_n, g^r x_k) \preceq m(g^r x_n, g^r x_{n+1})^s \cdot m(g^r x_{n+1}, g^r x_{n+2})^{s^2} \cdots m(g^r x_{k-1}, g^r x_k)^{s^{k-n}}$$

and similarly,

$$m(g^r y_n, g^r y_k) \preceq m(g^r y_n, g^r y_{n+1})^s \cdot m(g^r y_{n+1}, g^r y_{n+2})^{s^2} \cdots m(g^r y_{k-1}, g^r y_k)^{s^{k-n}}$$

Now since $sh < 1$ and $\delta_n \preceq \delta_0^{h^n}$, it follows from the two inequalities immediately above that

$$m(g^r x_n, g^r x_k) \cdot m(g^r y_n, g^r y_k) \preceq \delta_n^s \cdot \delta_{n+1}^{s^2} \cdots \delta_{k-1}^{s^{k-n}}$$
$$\preceq \delta_0^{sh^n(1 + sh + (sh)^2 + \cdots + (sh)^{k-n-1})}$$
$$\preceq \delta_0^{\frac{sh^n}{1-sh}} \to 1_E \quad as \quad n \to \infty$$

According to Lemma F.7, for any $c \in E$ with $1_E \ll c$, there exists $N_0 \in \mathbb{N}$ such that for any $n > N_0$, $\delta_0^{\frac{sh^n}{1-sh}} \ll c$. Thus it follows that for any $k > n > N_0$, Lemma F.7 implies $m(g^r x_n, g^r x_k) \cdot m(g^r y_n, g^r y_k) \ll c$, and so, $m(g^r x_n, g^r x_k) \ll c$ and $m(g^r y_n, g^r y_k) \ll c$. Consequently the sequences $\{g^r x_n\}$ and $\{g^r y_n\}$ are multiplicative Cauchy sequences in $g^r(X)$. Since $g^r(X)$ is multiplicative complete, there exists $x^\star$ and $y^\star$ in X such that $g^r x_n \to g^r x^\star$ and $g^r y_n \to g^r y^\star$ as $n \to \infty$. Since $\{g^r x_n\}$ is nondecreasing and $\{g^r y_n\}$ is non-increasing, notice from (c) and (d) of the Theorem, that $g^r x_n \sqsubseteq g^r x^\star$ and $g^r y^\star \sqsubseteq g^r y_n$. Now notice that,

$$m(F^r(x^\star, y^\star), g^r x^\star) \leq [m(F^r(x^\star, y^\star), g^r x_{n+1}) \cdot m(g^r x_{n+1}, g^r x^\star)]^s$$
$$= [m(F^r(x^\star, y^\star), F^r(x_n, y_n)) \cdot m(g^r x_{n+1}, g^r x^\star)]^s$$
$$\leq [m(g^r x^\star, g^r x_n) \cdot m(g^r y^\star, g^r y_n)]^{sM^\star q^r} \cdot m(g^r x_{n+1}, g^r x^\star)^s$$

Similarly, we have,

$$m(F^r(y^\star, x^\star), g^r y^\star) \leq [m(g^r y^\star, g^r y_n) \cdot m(g^r x^\star, g^r x_n)]^{sM^\star q^r} \cdot m(g^r y_{n+1}, g^r y^\star)^s$$

Now put $\tau = m(F^r(x^\star, y^\star), g^r x^\star) \cdot m(F^r(y^\star, x^\star), g^r y^\star)$, then it follows from the two inequalities immediately above that,

$$\tau \leq [m(g^r x^\star, g^r x_n) \cdot m(g^r y^\star, g^r y_n)]^{2sM^\star q^r} \cdot [m(g^r x_{n+1}, g^r x^\star) \cdot m(g^r y_{n+1}, g^r y^\star)]^s$$

CHAPTER 6. R-G-MONOTONE AND R-W-COMPATIBLE MAPPINGS IN ORDERED MULTIPLICATIVE CONE B-METRIC SPACE

Proof of Theorem F.1 Continued 1

Since, $g^r x_n \to g^r x^\star$ and $g^r y_n \to g^r y^\star$ as $n \to \infty$, then by Definition F.6, and for $1_E \ll c$, there exists $N_0 \in \mathbb{N}$ such that for all $n > N_0$, we have, $m(g^r x^\star, g^r x_n) \ll c^{\frac{1}{8sM^\star q^r}}$, $m(g^r y^\star, g^r y_n) \ll c^{\frac{1}{8sM^\star q^r}}$, $m(g^r x_{n+1}, g^r x^\star) \ll c^{\frac{1}{4s}}$, $m(g^r y_{n+1}, g^r y^\star) \ll c^{\frac{1}{4s}}$. Thus, it follows that $\tau \ll [c^{\frac{1}{8sM^\star q^r}} \cdot c^{\frac{1}{8sM^\star q^r}}]^{2sM^\star q^r} \cdot [c^{\frac{1}{4s}} \cdot c^{\frac{1}{4s}}]^s = c^{\frac{1}{4}} \cdot c^{\frac{1}{4}} \cdot c^{\frac{1}{4}} \cdot c^{\frac{1}{4}} = c$. Thus, according to Lemma F.7, $\tau = 1_E$, thus it follows that, $m(F^r(x^\star, y^\star), g^r x^\star) = 1_E$ and $m(F^r(y^\star, x^\star), g^r y^\star) = 1_E$, hence, $F^r(x^\star, y^\star) = g^r x^\star$ and $F^r(y^\star, x^\star) = g^r y^\star$, that is, $(x^\star, y^\star)$ is a coupled r-coincidence point of F and g

6.4 Exercises

Exercise F.1 1

Let the setting be this chapter. Deduce the higher-order version of Theorem 3.1 [Zaid Mohammed Fadail et.al, On mixed g-monotone and w-compatible mappings in ordered cone b-metric spaces, Math Sci (2015) 9:161–172]

Exercise F.2 1

(a) Show that Theorem F.1 is a special case of Exercise F.1

(b) Show that Theorem F.1 is the higher-order version of Corollary 3.2 [Zaid Mohammed Fadail et.al, On mixed g-monotone and w-compatible mappings in ordered cone b-metric spaces, Math Sci (2015) 9:161–172]

Exercise F.3 1

Let the setting be this chapter. Deduce the higher-order version of the following

(a) Corollary 3.3 [Zaid Mohammed Fadail et.al, On mixed g-monotone and w-compatible mappings in ordered cone b-metric spaces, Math Sci (2015) 9:161–172]

(b) Corollary 3.4 [Zaid Mohammed Fadail et.al, On mixed g-monotone and w-compatible mappings in ordered cone b-metric spaces, Math Sci (2015) 9:161–172]

(c) Theorem 3.5 [Zaid Mohammed Fadail et.al, On mixed g-monotone and w-compatible mappings in ordered cone b-metric spaces, Math Sci (2015) 9:161–172]

(d) Corollaries 3.6-3.8 [Zaid Mohammed Fadail et.al, On mixed g-monotone and w-compatible mappings in ordered cone b-metric spaces, Math Sci (2015) 9:161–172]

(e) Theorem 3.11 [Zaid Mohammed Fadail et.al, On mixed g-monotone and w-compatible mappings in ordered cone b-metric spaces, Math Sci (2015) 9:161–172]

(f) Corollaries 3.12-3.14 [Zaid Mohammed Fadail et.al, On mixed g-monotone and w-compatible mappings in ordered cone b-metric spaces, Math Sci (2015) 9:161–172]

Exercise F.4 1

Deduce that existence and uniqueness of a common coupled r-fixed point from Theorem F.1 will be a special case of Exercise F.3(c)

6.5 References

(1) Jungck, G, Radenovic, S, Radojevic, S, Rakocevic, V: Common fixed point theorems for weakly compatible pairs on cone metric spaces. Fixed Point Theory Appl. 2009, Article ID 643840 (2009)

(2) Bhaskar, T.G., Lakshmikantham, V.: Fixed point theorems in partially ordered metric spaces and applications. Nonlinear Anal. 65(7), 1379–1393 (2006)

(3) Lakshmikantham, V., Ciric, L.: Coupled fixed point theorems for nonlinear contractions in partially ordered metric spaces. Nonlinear Anal. 70(12), 4341–4349 (2009)

(4) Abbas, M., Khan, M.A., Radenovic , S.: Common coupled fixed point theorems in cone metric spaces for w-compatible mappings. Appl. Math. Comput. 217, 195–202 (2010)

(5) Ampadu, Clement (2016): r-Coincidence Point and r-Fixed Point Theorems in Multiplicative Cone b-Metric Space, Unpublished

(6) Zaid Mohammed Fadail et.al, On mixed g-monotone and w-compatible mappings in ordered cone b-metric spaces, Math Sci (2015) 9:161–172

www.ingramcontent.com/pod-product-compliance
Lightning Source LLC
Chambersburg PA
CBHW051103180526
45172CB00002B/753